Principles of Engineering

WORKBOOK

Principles of Engineering

WORKBOOK

Thomas Singer
PLTW Affiliate Professor
Sinclair Community College, OH

Teresa A. Phillips, BSE, MS

Debbie A. French
New Philadelphia City Schools, OH

Consulting Editor:

Pamela S. Lottero-Perdue, BME, ME.d., Ph.D.
Associate Professor of Science Education
Department of Physics, Astronomy & Geosciences
Towson University

Australia • Brazil • Japan • Korea • Mexico • Singapore • Spain • United Kingdom • United States

Principles of Engineering: Workbook
Thomas Singer, Teresa Phillips, Debbie French

Vice President, Career and Professional Editorial: Dave Garza

Director of Learning Solutions: Sandy Clark

Senior Acquisitions Editor: James DeVoe

Managing Editor: Larry Main

Product Manager: Mary Clyne

Editorial Assistant: Aviva Ariel

Vice President, Career and Professional Marketing: Jennifer McAvey

Marketing Director: Deborah S. Yarnell

Sr. Market Development Manager: Erin Brennan

Sr. Brand Manager: Kristin McNary

Production Director: Wendy Troeger

Production Manager: Mark Bernard

Content Project Manager: David Barnes

Art Director: Bethany Casey

Cover image: © Veer, A Corbis Corporation Brand/Ocean Photography

ISBN-13: 978-1-4354-2837-9

ISBN-10: 1-4354-2837-4

Delmar
5 Maxwell Drive
Clifton Park, NY 12065-2919
USA

Cengage Learning products are represented in Canada by Nelson Education, Ltd.

For your lifelong learning solutions, visit **delmar.cengage.com**

Visit our corporate website at **cengage.com**

Printed in the United States of America
8 9 10 11 12 13 14 25 24 23 22 21

Contents

Preface

This workbook was developed to support *Principles of Engineering* with real-world, hands-on activities that build basic skills for engineering and provide opportunities to apply those skills to more challenging projects. The authors have combined their years of experience teaching Project Lead The Way's® Principles of Engineering curriculum to produce a resource brimming with:

- Hands-on, directed engineering activities
- Drawing and sketching practice
- Math support
- Brainstorming and team development exercises
- Open-ended design problems and projects to provide greater challenges

Wherever necessary, sheets of blank engineer's notebook paper, as well as orthographic and isometric grids, are included for practice. As students complete the variety of activities in this workbook, they will be prompted to use these resources to develop their documentation skills.

Features of This Workbook

This text was developed to complement and support Project Lead The Way's® Principles of Engineering curriculum, and can be used to support any project-based course in engineering design. The following features are built into each unit to help students apply engineering principles to achieve productive results.

BACKGROUND

Background sections help students develop and review the knowledge they need to perform the activities that follow.

TIP SHEETS

Tip sheets alert students to common pitfalls and provide helpful hints and motivating anecdotes to smooth students' journey toward successful design.

EXERCISES

At the core of this workbook are dozens of hands-on exercises that build essential engineering skills from math to brainstorming to sketching, drawing, and team-building.

Principles of Engineering and Project Lead The Way, Inc.

This workbook is part of a series of learning solutions that resulted from a partnership forged between Delmar Cengage Learning and Project Lead The Way, Inc. in February 2006. As a nonprofit foundation that develops curricula for engineering, Project Lead The Way, Inc. provides students with the rigorous, relevant, reality-based knowledge that they need to pursue engineering or engineering technology programs in college.

Project Lead The Way® curriculum developers strive to make math and science relevant for students by building hands-on, real-world projects in each course. To support Project Lead The Way's® curriculum goals, and to support all teachers who want to develop project/problem-based programs in engineering and engineering technology, Delmar Cengage Learning is developing a complete series of texts to complement all of Project Lead The Way's® nine courses:

- Gateway to Technology
- Introduction to Engineering Design
- Principles of Engineering
- Digital Electronics
- Aerospace Engineering
- Biotechnical Engineering
- Civil Engineering and Architecture
- Computer Integrated Manufacturing
- Engineering Design and Development

To learn more about Project Lead The Way's® ongoing initiatives in middle school and high school, please visit www.pltw.org.

Acknowledgments

The authors and publisher wish to thank everyone who assisted in the development of this workbook. In particular, we are grateful to our consulting editor, Pamela Lottero-Perdue, Associate Professor of Science Education in the Department of Physics, Astronomy, and Geosciences at Towson University. Dr. Lottero's thorough and insightful reviews transformed the manuscript and increased the quality and pedagogical value of the activities.

In addition, we acknowledge Brett Handley for his substantial contribution to this workbook by providing content consultation, subject material, and original graphics.

To my wife, Anita, and daughter, Rachel, for providing me the support and space to accomplish this endeavor, thank you

—Tom

About the Authors

Thomas Singer is a Professor of Mechanical Engineering Technology at Sinclair Community College in Dayton, Ohio. He has written several educational texts on the Autodesk product tool. He also has worked extensively in virtual teaming and collaborative design tools. Mr. Singer is a Project Lead The Way affiliate professor for Introduction to Engineering Design, and co-authored the workbook *Engineering Design: An Introduction*. He is also a NISOD Teaching Excellence award winner.

Teresa Phillips works as a freelance curriculum writer. She has held positions with Project Lead The Way, Inc. as Director of Master Teachers and as Associate Director of Curriculum and Professional Development. Ms. Phillips earned a Bachelor of Science in Technology Education from the State University of New York at Oswego and a Master of Science in Cross-Disciplinary Professional Studies from the Rochester Institute of Technology, with concentrations in engineering-technology education. She holds a New York State permanent teaching certification and has been a master teacher in Computer-Integrated Manufacturing and Introduction to Engineering Design.

Debbie French teaches Principles of Engineering, Physics, and Physical Science at New Philadelphia High School in New Philadelphia, Ohio. She holds a Bachelor of Arts from Denison University and a Master of Arts in Teaching from Miami University.

Ms. French serves as co–Principal Investigator of the National Science Foundation (NSF) grant, "Exploring Innovative STEM Education Through Guitar Design and Manufacture." This grant trains high school and community college faculty on how to build a solid-body electric guitar and implement guitar-themed STEM activities.

She is also a member of the NASA/IPAC Teacher Archive Research Program (NITARP) 2012 class and is working with a group of high school, community college, and informal educators with Dr. Luisa Rebull, of Caltech, on identifying young stellar objects in BRC 27.

Principles of Engineering

WORKBOOK

CHAPTER 1
Overview and History of Engineering

Before You Begin

Think about these questions as you study the concepts in this chapter.

- When did engineering begin?
- What were some of the first engineering designs?
- Who were the important pioneers in the field of engineering, and what did they contribute?
- What are the first steps to becoming an engineer today?
- How have advances in technology spurred growth in engineering fields?

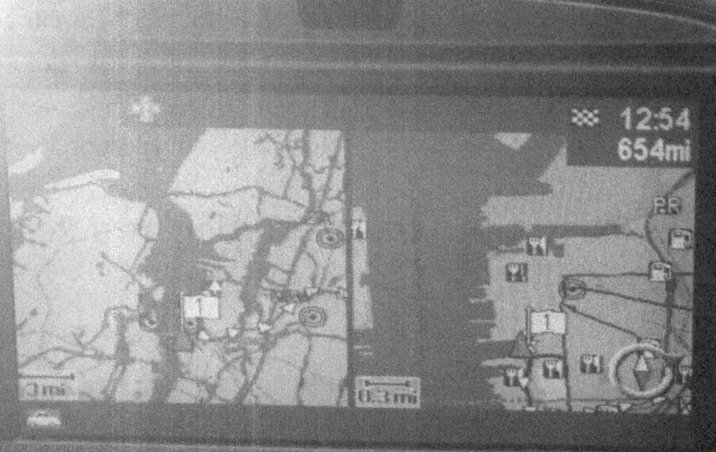

Exercise 1.1 Engineering Timelines

Objective

Determine how historical developments apply to engineering. Many of the major time periods have been influenced by engineering discoveries.

Procedure

1. Break into teams.
2. As a team, select a 200- to 500-year time period in history.
3. As a team, identify one engineering discovery or feat that was accomplished during your chosen time period.
4. Document the location of the engineering discovery or feat, including any persons involved in the discovery. Be sure to document your research sources. Present your research to the class.
5. Class discussion question: Based on the class presentations, what can we say about the rate of innovations, discoveries, and feats of engineering over time?

Exercise 1.2 Companies in Our Community

Objective

Research the heritage of your community. Knowing this information will help you understand the reasons for settling in the area and what has driven the growth and change that has occurred locally. Each region of the United States has at least one example of an industry that was or still is an important factor in that region's history. Examples of companies that have succeded are Coca-Cola™ and McDonalds™ shown in figure 1-1. Examples include tire manufacturing in Akron, Ohio, and textile manufacturing in South Carolina.

Procedure

1. Research the industry or company history: How did the company start?
2. Make a time line of the company, showing major improvements and failures. Examples include the Coca-Cola® Company's introduction of "New Coke" and the subsequent return of the original formula, now known as Coca-Cola Classic®, or the rise and fall of the McDLT® hamburger.
3. If the company is no longer in business or has relocated, what caused the downfall or relocation?
4. Create a five-minute presentation and report on your findings.

© iStockphoto/nico_blue

FIGURE 1-1 *Two companies that have tested new products. Some have made it; others have not.*

Exercise 1.3 Transportation as a Resource

Objective

Transportation is a key aspect of all societies. Determine the linkage between historical transportation systems and current transport systems.

Procedure

In teams of two, select a current mode of transportation. In a report,

1. Explain the method of propulsion.
2. Determine the current cost of operation (per mile or kilometer).

3. Describe the underlying technology on which this system was built. (Example: Turbine locomotives are built on the technology behind the 18th-century steam locomotive.)
4. Project what the future holds for the mode of transportation that your team selected.

Example topics include trains, planes, automobiles, bicycles, rickshaws, space vehicles, submarines, and boats. Your report should not fill more than two written pages.

Exercise 1.4 Careers in Engineering

Objective

Engineering can lead to a variety of career choices. Most people have heard of mechanical, industrial, aeronautical, or electrical engineering. However, many new specialties are now emerging in the broad field of engineering. Explore some of the more exotic pathways to an engineering career. You will find that these specialties are subsets of the more general engineering areas.

Procedure

Research and describe in a written report the most exotic engineering career that you can find. Your write-up should be one page using 12 point type and 1.5 line spacing, and it should include up to two pictures that measure no more than 2" × 2" square.

Exercise 1.5 Energy as a Resource

Objective

Energy is the engine that expands civilizations. The development and exploration of existing and new energy resources are important tasks. Identify and categorize the agencies that coordinate, manage, or set standards for energy use.

Procedure

The generation and use of energy by industry and society is an ever-increasing concern among the human population. Choose an engineering activity, and identify a coordinating energy-efficiency organization. Examples include Energy Star for appliance manufacturing and the Leadership in Energy and Environmental Design (LEED) (figure 1-2) for construction engineering.

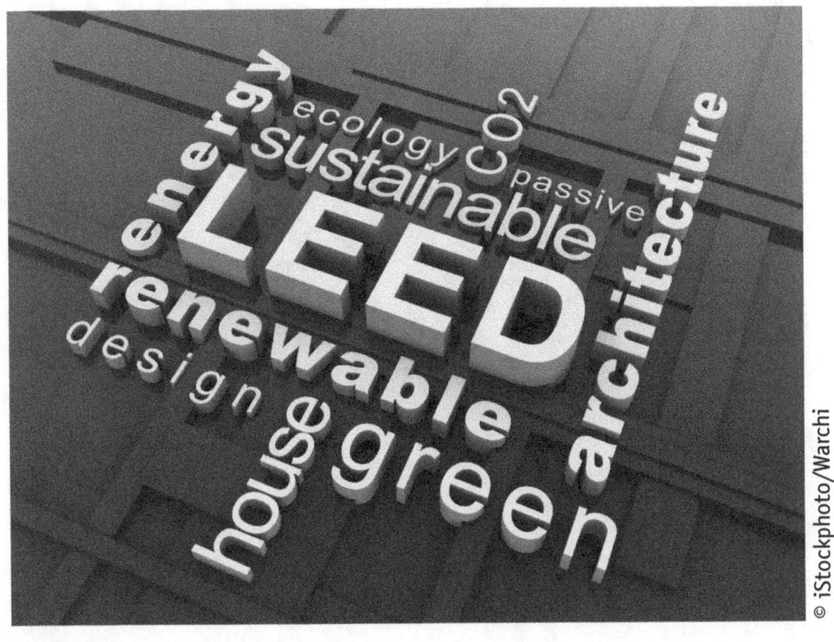

FIGURE 1-2 *LEED certification and what it means.*

Coordinating Organization	Engineering Activity	Standards or Regulations	Engineering Careers Affected
Energy Star	Appliance manufacturing	Measures energy usage, sets efficiency standards that appliances are measured against	• Manufacturing engineers • Mechanical engineers • Packaging engineers • Electrical engineers

TABLE 1-1 *A sample chart that identifies and categorizes the agencies that coordinate, manage, or set standards for energy use.*

As a class, develop a chart that indicates the coordinating organization, the engineering activity addressed by each organization, the types of regulations or guidelines that the organization creates, and the engineering careers linked to the products or processes listed. A sample chart is illustrated in Table 1-1.

CHAPTER 2
Design Tools

Before You Begin

Think about these questions as you study the concepts in this chapter.

- How does the engineering process guide the engineer from a loosely understood problem to a highly refined solution?
- What brainstorming techniques do engineers use to generate ideas during a design process?
- What kinds of sketches do engineers use to record their initial ideas?
- What kinds of computer-based modeling methods do engineers use to develop their solutions?
- How and why do engineers use analytical tools?
- How do engineers communicate and present their ideas?

Exercise 2.1 Defining the Design Process

Objective

Identify the steps of the design process.

Procedure

1. Using Figure 2-1, label the design process steps in order.
2. Provide a written statement of what occurs during each step of the design process.
3. Search for and provide drawings or pictures that illustrate each step of the design process. Develop a Microsoft® PowerPoint® presentation listing each design step and its corresponding image. Be sure to document your image sources.

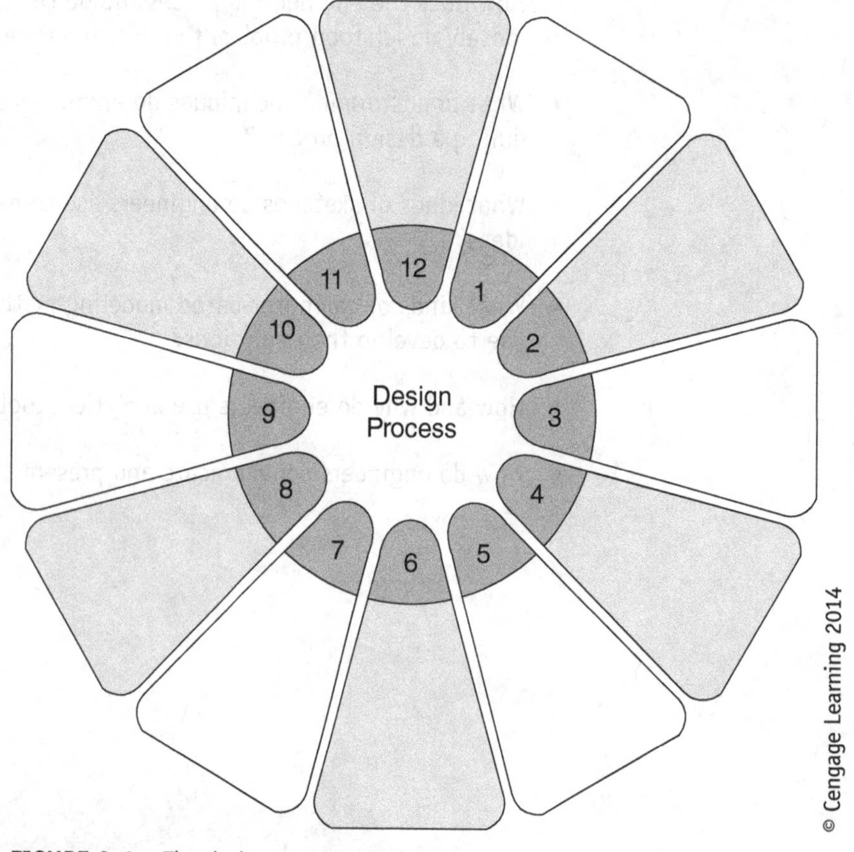

© Cengage Learning 2014

FIGURE 2-1 *The design process.*

Exercise 2.2 Applying the Design Process

Objective

Applying the design process to an existing product provides a look at how the product has come to market. It also provides the level of detail needed for a designer.

Procedure

1. Choose a manufactured product that you are familiar with (for example, cell phone, camera, iPod™, gaming system, cosmetics, shoes, etc.)

2. Study the 12-step design process and sketch (in words or pictures) how the design process brought the product that you are studying to the marketplace. You should have 12 unique answers, one for each step.
3. Determine the challenges that the designers faced in manufacturing the product and bringing it to market (refining the solution).
4. Describe how you would improve the product's design or manufacture.
5. Indicate the hardest step that you encountered in the 12-step process, and describe why it was your hardest step during this project.

Sketching and Design Solutions

Drawing and sketching objects and ideas dates back to cave paintings in prehistoric times. The drawing has always been used as a key form of communication to construct and build objects. That is no different today; drawings and sketches are a required aspect of proper object creation. This section will help you develop the skills needed for good sketching and drawing communication.

Exercise 2.3 Sketching Practice

Objective

Being able to sketch items accurately is an important skill for an engineer. The sketch is the tool that conveys the story of the object. Select three small objects within your environment and create orthographic and isometric sketches of these objects.

Materials

3 sheets of orthographic grid paper

1–3 sheets of isometric grid paper (a single sheet can be quartered for this project)

Procedure

1. Find three objects in your classroom or at home that can be held in your hand.
2. Develop orthographic sketches (top, front, right side) for each object.
   a. Select the primary side of the object and use that as the front view.
   b. Project the side view and the top of the object (if needed).
   c. Add general dimensions of the object for reference.
3. Using the 30°, 150°, and 90° technique, develop an isometric sketch of the objects that you selected.

Continued from page

Continued to Page

SIGNATURE:		DATE:
WITNESSED BY:	DATE:	PROPRIETARY INFORMATION

Continued from page

Continued to Page

SIGNATURE:		DATE:
WITNESSED BY:	DATE:	PROPRIETARY INFORMATION

Continued from page

Continued to Page

SIGNATURE:		DATE:
WITNESSED BY:	DATE:	PROPRIETARY INFORMATION

Exercise 2.4 Isometric Sketching

Objective

Isometric sketching provides a visual perspective that is used to help people who cannot visualize an object. Many customers or clients need that visual cue to help them understand the design. Develop an isometric drawing with callouts identifying different parts of an object.

Procedure

1. From your kitchen at home, choose a utensil that has two or more parts (for example, a potato peeler, spatula, or ice cream scoop).
2. Develop an isometric drawing with callouts identifying the different parts of the object. Include fasteners such as screws, rivets, and welds in your callouts.

Exercise 2.5 Brainstorming

Objective

Brainstorming is the best way to get a creative direction for a development. Document all your ideas because an idea that isn't used in the current project may work better in the next.

Procedure

1. Break into teams of four. As a team, select one of the design challenges listed below.
2. Conduct a brainstorming session to identify possible solutions.
3. Create a chart of core criteria items (cost, form, fit, function, etc.) to evaluate the brainstorm designs.
4. Develop a criteria matrix that will be used to develop a decision matrix, as illustrated in Table 2-1.
5. Develop a brainstorm evaluation matrix. The matrix should be based on the core criteria items that your team has decided upon.
6. Grade each solution to the problem (for a minimum of 10 total solutions), as illustrated in Table 2-2.

Criteria	Brainstorming Solution	Description of the Criteria	Observation of the Solution
Guitar material	Wood-hickory	Material must be easy to manufacture.	Hickory has a very high material hardness scale, so it is difficult to CNC-machine.

TABLE 2-1 *Sample Criteria Matrix.*

Criteria Guitar material	Solution to Criteria	Description	Evaluation: 1 = Poor, does not meet criteria 5 = Excellent, exceeds criteria					
			1	2	3	4	5	
	Hickory	Wood has a very high hardness factor.	X					
	Maple	Wood has a medium hardness factor – has been used to make other guitars.				X		
Design Challenge 1: Guitar Design								
You are developing a guitar for the lead guitarist for your favorite music group (as decided by the team). This guitar must embrace the group's genre and style. Brainstorm potential designs for the guitar.								
Design Challenge 2: Cell Phone Design								
A new cell phone manufacturer wants to break into the market with a phone design specifically for teenagers. What should this phone look like, and what features should it have?								
Design Challenge 3: Locker Organizer								
Students sometimes forget their homework and binders in their lockers, both when they head to class and when they head home for the day. Their forgetfulness causes late papers and lower grades. Devise a solution that will help students remember their work. Any solution is fair game, from binders to notebooks to cell phones and beyond.								
Design Challenge 4: Make Your Own Problem								
As a team, identify and define a problem, and develop solutions in a brainstorming session.								

TABLE 2-2 *Sample Evaluation Matrix.*

CAD

Creating computer-aided design (CAD) models is an important step in the design-to-manufacturing process. CAD models provide the ability to see the design, but they also allow you to apply analysis tools to help make decisions on the product/object.

Exercise 2.6 CAD Model Development

Objective

Create CAD models of the sketches that you developed as part of the sketching activity.

Procedure

1. Choose one of the sketches that you completed in Exercise 2.3, Exercise 2.4, or Exercise 2.5.
2. Model your sketch in a CAD system. You might need to do some reverse engineering to find the dimensions of items.

Exercise 2.7 Reverse Engineering of Parts

Objective

Take readily accessible items and develop them as CAD models to help create a database of tools that can be used later in the class.

Procedure

Select different Kinetix™ or VEX™ parts. Reverse engineer and develop the CAD models of the selected parts. All part models are then provided in a database for the class to use on future projects for designs using the Kinetix or VEX parts. Create a full-scale design of the model.

Exercise 2.8 CAD Product Search

Objective

A variety of CAD software tools are commercially available to develop 3-D models. In this exercise, identify two CAD software products that are used in the development of mechanical and architectural applications.

Procedure

1. Split up into teams of two.
2. CAD software products come from many different manufacturers and are designed for both general and specific industries.
3. Develop a matrix of the CAD product tools that you have discovered through research. A sample is illustrated in Table 2-3.

CAD product	Manufacturer	Field Used	Platform (PC, MAC, Xbox, PS3, Phone)	Cost
Aspire	Vectrix	Woodworking	PC	$795

TABLE 2-3 *A Sample Matrix for CAD Product Tools.*

4. As a class, develop a chart listing the CAD packages that each team found and determine the average number of packages found in the class.

Grading is based on the following criteria:

- Team with the most CAD packages: 100 points
- Teams above the average number of packages found: Range from 80–90 points
- Teams at the average number of packages found: 75 points
- Teams below the average number of packages found: 65–75 points
- Teams that did not try: 0 points

Explore and Select an Approach

It is important to realize that there is more than one solution to a design problem. The design process involves brainstorming and choosing an approach that will result in a workable solution. Figure 2-2 shows a team involved in the process of developing a solution.

FIGURE 2-2 *Norming a team decision.*

Storming, Norming, and Documenting Decisions

All teams that are formed go through a cycle of evolution that has specific stages. The timetable for each stage will vary with the team and the individual personalities of the team members. Being aware of these stages will make you a more effective team member and help move the team faster into a productive environment.

Meeting The first project meeting often introduces the members of a team to each other. Each member brings a unique role to the team and an agenda or set of personal goals.

Storming Team meetings often involve brainstorming, or the free flow of ideas that could lead to a design solution. Many times, the first idea is not the final one; it gets improvedover the course of the meeting.

Norming Once the group has finished brainstorming, one idea will emerge as a potential solution, or several ideas might morph into a single solution to the original problem. This process is called "norming." It is a key step toward ensuring that the group agrees on the final solution.

Resolution Resolution, the final step, formalizes the solution that the group has agreed upon. Until it this solution is documented, it really does not exist.

Documenting Documenting your team's solution is an important step. Teams can document their solutions in a variety of ways. The goal is to communicate ideas clearly to the key decision-makers—customers, financiers, the public, the government, etc. The form of communication can be written, verbal, electronic, media presentations (such as video), drawings, spreadsheets, or charts.

Team Management Best practices for team management include *listening* to the team members. Each member brings valuable information to the decision making process.

Exercise 2.9 How Are Decisions Made?

Objective

How do people make decisions? Research and find out what makes people select and decide. Making a decision is a complex process. What affects a decision? Why is one selection better than another?

Procedure

Answer the following question: How does a person make a decision?

Research how decisions are made using several websites like Mindtools.com. Write a two-page paper (12 point font, 1.5 line spacing) on how decisions are made. The last paragraph needs to explain what influences you in making decisions.

Exercise 2.10 The Decision Matrix

Objective

Develop a decision matrix for a design brief. The ability to focus on key factors in a design brief or client conversation is very important to meeting the needs of the client.

Procedure

Develop a decision matrix for the following design brief:

Design Brief	
Client:	Office supply megastore
Designer:	Matt M.
Problem Statement:	Push pins seem to expand in their storage space and fall out of the current desktop holding tray, or cause users to get poked when they try to select one.
Design Statement:	Design a unique push pin–holding system for a desktop.
Constraints:	1. 2-week design deadline 2. Cost to customer cannot exceed $8.50 3. Must prevent pokes to consumers using push pins 4. Must hold 50 push pins 5. Must be made of a renewable resource material

Exercise 2.11 Engineering Team Role-Play

Objective

Using the concepts in this chapter, create a team representing the diverse roles described below. Work together to form a solution based on the information provided in the design brief. Areas of concentration include teaming concepts, research of engineering standards and similar solutions, and aspects of governmental interactions.

Procedure

Please refer to the instructions below and the design brief to complete this exercise.

1. Break up into teams of 8 to 10 students.
2. Your teacher will provide slips with job descriptions for the following team roles. Each student will blindly draw a job description.
   a. (up to 2) Electrical engineer, utility company—Responsible for hooking up the wind farm to the electrical grid
   b. (up to 2) Concerned citizen(s)—Has read a lot about wind farm noise, bird disruption, limited output, and other issues
   c. Wind farm managing engineer—In charge of the wind farm project (addresses all wind farm technical questions)

d. Wind farm structural engineer—Focuses on the structure of the wind turbine
e. Wind farm electrical engineer—Determines electrical connections and interfaces
f. Local government official—Manages the county development
g. Weather specialist on wind patterns
h. Civil engineer—Works as consultant to wind farm group for site location

You will have a 45- to 60-minute period to complete the initial brainstorming session. Each group will then reveal their findings to the rest of the class in the form of an oral presentation.

Use the following design brief to understand your criteria and constraints.

Wind Turbine Farm Development Design Brief
Problem Statement: The West Texas town of Texhomex has been identified as a good area for the development of a wind turbine farm. The farm needs to be located within 35 miles (56 kilometers) of the center of the town, and 40 1.65-megawatt (MW) wind turbines will need to be constructed.
Design Statement: You will need to decide where to place high-voltage feeder lines for access to the wind turbine site and how far apart the wind turbines need to be placed. You also need to determine whether there are access roads for maintenance, and what the local and state laws are on wind turbines. There may also be federal regulations from the U.S. Department of Energy that you will need to research. Blue Canyon Wind Farm, which is a functioning wind farm project in Texas, is a good location for additional information on wind farm turbine placements and scope on this type of project.

FIGURE 2-3 *Wind turbines over hay bales in West Texas.*

Materials

- A computer with Internet access
- A printer (a color printer is preferred for map development)
- Screen capture and photo editing software

Wind Turbine Team Roles

1. **Electrical engineer, utility company (up to 2)**
   a. **Role:** Responsible for hooking up the wind farm to the electrical grid.
   b. **Action:** The role of these team members is to research and determine the location of high-tension electrical lines. By using Google maps and other mapping technology and researching local power companies, they should be able to determine where they are located.
2. **Concerned citizen (up to 2)**
   a. **Role:** These team members are worried about the location of the wind farm in proximity to any farmhouses or towns.
   b. **Action:** Research wind farm noise and wildlife and safety issues.
3. **Wind farm managing engineer**
   a. **Role:** This person is in charge of the wind farm project and addresses all wind farm technical questions.
   b. **Action:** Research wind farms and the positive and negative aspects of using wind turbines for energy generation
4. **Wind farm structural engineer**
   a. **Role:** This person needs to address how wind farms are laid out and placed (primarily on ridges) and what type of infrastructure is needed (foundations, electricity, water, etc.).
   b. **Action:** Research how a wind turbine is secured to the ground and what infrastructure is needed to make a turbine operate.
5. **Wind farm electrical engineer:**
   a. **Role:** This person researches the efficiency aspects of wind farms and addresses the managing engineer and concerned citizen's requests. This person also must work with the utility engineers to get the wind farm on the grid.
   b. **Action:** Locate the high-tension power lines closest to the project location and map a path for the distribution system.
6. **Local government official**
   a. **Role:** This person manages the county development and needs to research and study state and local regulation on wind farms.
   b. **Action:** Research the county and state building regulations (and federal regulations, as applicable) on wind turbines.
7. **Weather specialist on wind patterns**
   a. **Role:** This person uses the National Weather Service to research wind patterns, lightning issues, and other elements. His or her findings will need to be reported to the structural engineer making decisions on the structure of the towers.
   b. **Action:** Research the wind speed fields of the project areas.
8. **Civil engineer**
   a. **Role:** This person works as a consultant to a wind farm group for site location. Duties include researching wind farms, specifically the layout patterns and access issues, and working with the structural, local government, weather, concerned citizens, utility company, and managing engineer to draw up an acceptable layout.
   b. **Action:** Interpret, as an outside voice, what the wind farm team has brought to the design, and create the presentation that will be shown to the public.

Conclusion

The questions that each team will need to answer are as follows:

1. What plot of land did you choose for the wind farm?
   a. Why did you choose this land? (Provide a map location.)
   b. Are there high voltage power lines nearby that can be connected to the power grid?
   c. Are there access roads to the land area?
2. What are the local regulations in the Texhomex area regarding wind turbines?
3. Are there any structural or geological issues with placing turbines in this area?
4. What is the approximate cost of placing a wind turbine? How much would placing 40 of them cost?

Exercise 2.12 Rapid Prototyping

Objective

Rapid prototyping is a process developed in the mid-1990s to allow a concept design to be developed into a part and put into a designer's hands in just a day or so. There are many different prototypes that can be used. This exercise explores the prototyping options.

Procedure

1. Identify the various prototyping styles, including both additive and subtractive styles.
2. Choose a prototyping style, and track down a company that manufactures that type of prototyping.
3. Split into development teams. Find a local company that does prototyping for industrial clients and research what the benefits of prototyping are.
4. Have each team select a prototype method and create a four-slide PowerPoint presentation to "sell" why its prototyping method is better than the others.
5. Have the class vote on the most persuasive presentation.

Exercise 2.13 Online Prototyping Cost Analysis

Objective

Costs for prototyped parts vary greatly based on the process and materials. Determine the cost of prototyping a CAD model.

Procedure

Develop a 3-D part with CAD design software. Using an email account set up for class inquiries, send your part to an online rapid prototyping quoting website. Have them quote the design using different rapid prototyping processes.

1. Develop a spreadsheet chart to list the processes and costs, as well as the time needed.
2. Why is there a cost difference in the processes?
3. Select one of the processes and develop a justification sheet why your project manager (teacher) should fund the development of your prototype piece.

Using Analysis Tools

Spreadsheets are used to qualify and organize the data you collect. They can also provide analysis on the data to help guide you to a solution.

Exercise 2.14 Spreadsheet Profit Calculation

Objective

Use a spreadsheet to calculate profits.

Procedure

1. Study the following list of income and expenses to operate a 10-minute oil-change business. Construct a spreadsheet that will calculate the business's cost per oil change.
   - Price per standard oil change (5 quarts of oil): $27.50
   - Technician's wage: $10 per hour
   - Oil: $1.50 per quart
   - Filter: $2.50 per filter
   - Additional fluids filled: $1.50 per car
   - Supplies (rags): $1 per car
   - Oil reclamation fee: $1.50 per car
2. If the business employs one technician to service four cars per hour, what are the profits (income minus expenses) in one eight-hour workday? In one six-day workweek?
3. If the technician's wage increases to $12 per hour, what is the difference in profits for one day? For one week?

Exercise 2.15 Spreadsheet Sorting and Highlighting

Objective

Develop a sorting structure based on an existing spreadsheet.

Procedure

Develop a spreadsheet using the values illustrated in Figure 2-4. (Your teacher may provide you with an existing spreadsheet.) Using the Sort tool, determine which dealer has the best car purchase deal, based on the calculation of sales price versus the invoice price of the car. In addition, you may use the Conditional Formatting tools to shade any percentage below 80%, 90%, and 95% of the invoice price.

Dealer	Price	Fees	Tax + license	Price with tax title	Invoice		% of invoice
Dealer A	$25,957.49		$1,787.24	$27,744.73	$26,633.00		
Dealer B	$24,827.59	$250.00	$1,713.79	$26,791.38	$26,633.00		
Dealer C	$25,417.00		$1,752.11	$27,169.11	$26,633.00		
Dealer D							
Dealer E	$25,058.00		$1,728.77	$26,786.77	$26,633.00		
Dealer F					$26,633.00		
Dealer South A	$24,533.00		$1,694.65	$26,227.65	$26,633.00		
Dealer South B	$25,470.00		$1,755.55	$27,225.55	$26,633.00		
Dealer South C	$23,903.00	$399.00	$1,653.70	$25,955.70	$26,633.00		
Dealer South D	$25,709.00	$599.00	$1,771.09	$28,079.09	$26,633.00		

© Cengage Learning 2014

FIGURE 2-4 *A sample spreadsheet.*

Exercise 2.16 Vehicle Leasing vs. Buying

Objective

Determine if buying a car is better than leasing a car from a financial perspective.

Procedure

Find a car that you would like to drive. Research the price of the car (websites have dealer invoicing and typical prices). Using that data, develop a spreadsheet to determine how much the car will cost if it is owned for 3 years. This needs to be done as an outright purchase (with and without a loan) and as a lease of the vehicle. For the purchase with a loan, use 10% of the purchase price as a down payment in your cost calculations. Loan interest rates are based on your financial credit history, known as a credit score. These scores will vary from a low of 400 to a high of 900. For this project, assume a credit score of 750, which will get you the best interest rates available for both the purchase and lease. These interest rates can be found at local banks, dealerships and online.

Bonus Task: Figure out the same problem for a 6-year period. Using the same vehicle for the purchase for the whole 6 years and two lease vehicles (one from year 1–3 and the second at a 10% cost increase for years 4–6). (Leases are not written for a 6-year period, so you must lease two cars for a 3-year period apiece instead.).

Double Bonus Task: Estimate the service costs of the vehicles based on typical service over the 6 years for the car (or the two cars, in the case of leasing) and use that data to help determine the best option: purchasing or leasing?

Design Problems

The following design problems can be done in teams of two, or people can work on them as individuals. Each problem may have unique deliverables. If projects are done individually, then after the brainstorming session, have another person (whether another student or a family member) help refine the resulting ideas based on the standard criteria and sign off on the list to replicate a customer sign off.

Exercise 2.17 Soft Drink Bottle Design

Objective

Develop new soft drink bottle designs.

Procedure

A soft drink manufacturer needs a new bottle design concept for a specific product line targeted to teenagers. This design change must be used for all container types (12-oz. cans, 16.9- and 20-oz. bottles, and 2-liter bottles). The marketing campaign must incorporate the outside activities that teens love to do.

Remember that the materials are different among the containers, so make a design that can be replicated in all existing material types and also can accommodate the use of a new material in the future.

Sketch a concept design for each of the three classes (cans, small bottles, and large bottles) of containers, explain how each design links to the theme, and provide a general color scheme of the containers.

Exercise 2.18 Power Cord Management

Objective

Develop a power cord management solution.

Procedure

Cell phones, cameras, and MP3 players all have unique cords and plugs. Some manufacturers use standardized plugs, but others do not. Most people just throw the cords and chargers in a drawer or leave them lying around, inviting the potential for loss or damage. Use the design process to develop a new way to manage cords and chargers. You can create an explanation, draw a sketch, or both.

Exercise 2.19 Luggage Scale

Objective

Develop a tool that will provide the accurate weight of a piece of luggage.

Procedure

Airlines now charge luggage fees for travelers with checked bags. They also limit the amount of weight you have in your checked bag to 50 pounds; if your luggage weighs more than that limit, you pay an additional fee.

Design a tool that can be used on any piece of luggage to weigh it accurately. The object needs to be small enough and light enough not to add significant weight to any bag if it is left attached or if it is packed. Develop a storyboard, including sketches and text, that outlines the features and look of your design.

Continued from page

Continued to Page

SIGNATURE:		DATE:
WITNESSED BY:	DATE:	PROPRIETARY INFORMATION

Continued from page

Continued to Page

SIGNATURE:		DATE:
WITNESSED BY:	DATE:	PROPRIETARY INFORMATION

Continued from page

Continued to Page

SIGNATURE:		DATE:
WITNESSED BY:	DATE:	PROPRIETARY INFORMATION

Continued from page

Continued to Page

SIGNATURE:		DATE:
WITNESSED BY:	DATE:	PROPRIETARY INFORMATION

CHAPTER 3
The Mechanical Advantage

Before You Begin

Think about these questions as you study the concepts in this chapter.

- What is force, and how is it applied?
- What is the difference between work and power?
- What is mechanical advantage, and how is it a trade-off between the effort and load with respect to force and distance?
- What is the difference between a machine's actual mechanical advantage (AMA) and its ideal mechanical advantage (IMA), and why is one always smaller in magnitude than the other?
- What are the six simple machines, and how do you calculate their mechanical advantage?
- What is a compound machine, and how do you calculate its ideal mechanical advantage (IMA)?

BACKGROUND

Force, Work, and Power

If you were going to push a bed across the floor of your room, chances are that you would instinctively lower your body so that your effort is more in line with the direction that you want the bed to move.

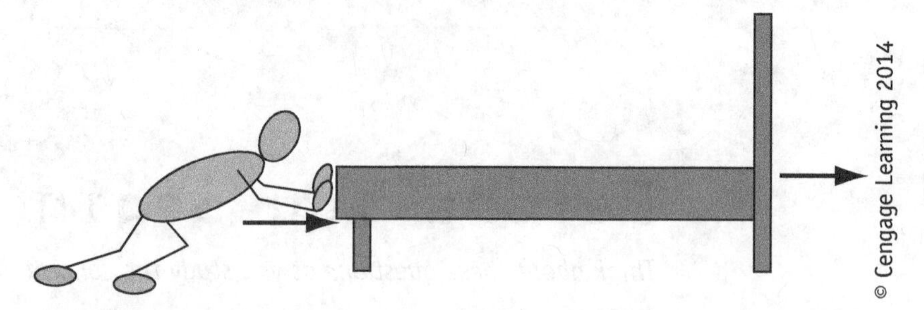

FIGURE 3-1A *Applied force in line with a desired direction.*

That is because you intuitively know that it is easier to move the bed if the applied force is parallel to the desired direction of movement and directly opposing friction (Figure 3-1A). Likewise, you also intuitively know that it would be more difficult to move the bed if the force were applied at a downward angle from a standing position (Figure 3-1B).

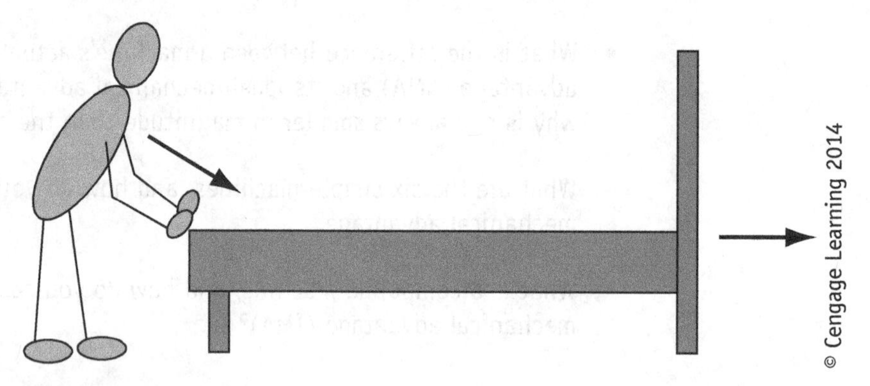

FIGURE 3-1B *Applied force at a downward angle to the desired direction.*

Recall that force is a vector quantity, which means that it has both magnitude (answering the question: "How much?") *and* direction (answering the question "Which way?"). When a person applies a force at an angle, as shown in Figure 3-1B, the force can be separated into vertical and horizontal component forces. The horizontal component, F_x, pushes against friction and helps to move the bed along the floor. The vertical component, F_y, simply pushes downward on the floor (in fact, increasing the amount of friction that the person must push against!) See Figure 3-1C for a depiction of the force, F, and the component forces, F_x and F_y.

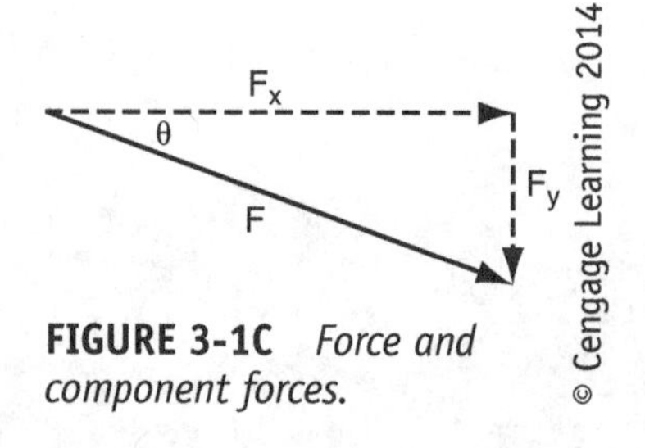

FIGURE 3-1C *Force and component forces.*

Only the horizontal force, F_x, contributes to the work, W, done by the person on the bed to move the bed horizontally.

Similar to the case of the person pushing the bed, in Exercise 3-1, only the horizontal component, F_x, of the force vector (which is parallel to the direction of motion) affects the movement of the box. To find the value of the X-component force vector, the trigonometric equation for the cosine of an angle must be used:

$$\cos\theta = \frac{\text{adjacent}}{\text{hypotenuse}}$$

$$\cos\theta = \frac{F_x}{F}$$

$$F_x = F\cos\theta$$

Exercise 3.1 Calculating Work and Power

Objective

At the conclusion of this exercise, you will be able to do the following:

1. Apply trigonometry to determine the X-component of a force vector.
2. Calculate the amount of work being performed on an object.
3. Calculate power output as a function of work and time.

Procedure

Read the section on force, work, and power in Chapter 3, "The Mechanical Advantage," in the *Principles of Engineering* textbook.

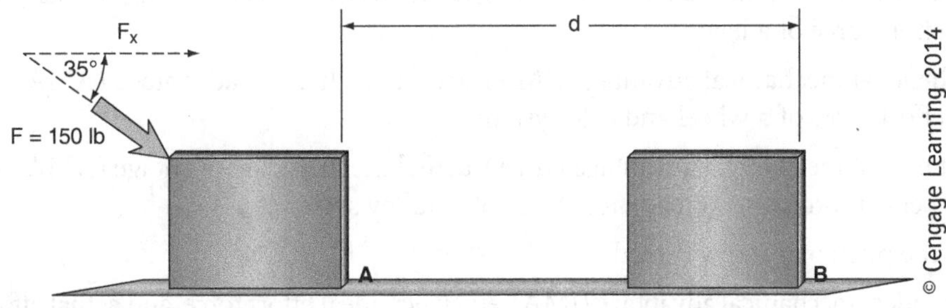

FIGURE 3-2 *Work.*

1. In Figure 3-2, a constant 150-lb force is applied to a box at a 35° angle relative to the horizontal floor, pushing against friction and sliding the box 8 feet across the floor from point A to point B in 10 seconds. Use this information to answer the following questions. Show your math work in the spaces provided.

   a. What is the value of the horizontal component of the force vector?

   F_x = ______________________________

   b. How much work was done to the box?

   W = ______________________________

   c. How much power was exerted, in watts?

   P = ______________________________

Simple Machines

Explore Your World

Locate four different kitchen, gardening, or other tools that use one or more simple machines. For each tool, create a pictorial sketch of the tool that is annotated with:

- Overall dimensions (e.g., overall length and width) of the tool
- A description of the operation of the device
- The type(s) of simple machine(s) that the tool uses

See Figure 3-43 in your textbook for an example of an annotated sketch.

Exercise 3.2 Calculating Mechanical Advantage

Objective

At the conclusion of this exercise, you will be able to do the following:

1. Identify the six simple machines.
2. Calculate the slope, ideal mechanical advantage (IMA), actual mechanical advantage (AMA), efficiency, ideal effort force, and actual effort force of an inclined plane.
3. Calculate the slope, ideal mechanical advantage (IMA), actual mechanical advantage (AMA), load capacity, and ideal effort force of a wedge.
4. Identify the three classes of levers.
5. Calculate the ideal mechanical advantage (IMA), actual mechanical advantage (AMA), ideal effort force, and actual effort force of a lever.
6. Calculate the ideal mechanical advantage (IMA), actual mechanical advantage (AMA), ideal effort force, and actual effort force of a wheel and axle system.
7. Calculate the ideal mechanical advantage (IMA), actual mechanical advantage (AMA), effort rope length, ideal effort force, and actual effort force of a pulley system.
8. Determine the pitch of a screw thread.
9. Calculate the ideal mechanical advantage (IMA), efficiency, ideal effort force, and actual effort force of a screw.

Procedure

Read the section on mechanical advantage in Chapter 3, "The Mechanical Advantage" in the *Principles of Engineering* textbook.

Problem 3.1 The inclined plane shown in Figure 3-3 is used to move load L a vertical distance h. The actual amount of effort force that is required to move the load up the inclined surface is 50 lb. Complete the following exercises using the values L = 100 lb; $\theta = 25°$, and h = 42 in. Show your math work in the spaces provided.

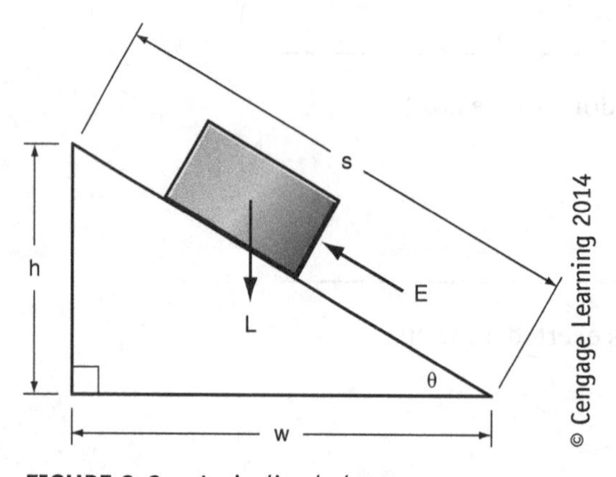

FIGURE 3-3 *An inclined plane.*

1. Calculate the length, *s*, of the inclined plane's slope.

   s = ______________________________

2. Calculate the ideal mechanical advantage (IMA) of the inclined plane.

   IMA = ______________________________

3. Calculate the actual mechanical advantage (AMA) of the inclined plane.

AMA = ______________________________

4. What is the efficiency (η) of this inclined plane system?

η = ______________________________

5. What is the ideal effort force (E_I) that would be needed to push the load up the inclined plane in the absence of friction?

E_I = ______________________________

Problem 3.2 The table shown in Figure 3-4 contains data that describe four inclined planes, A, B, C, and D. Use the information provided to calculate the missing data for each of the four inclined planes. Show your math work in the spaces provided.

	Load Weight (lb)	Slope Length (in)	Inclined Plane Height (in)	Inclined Plane Width (in)	Slope Angle (degrees)	E_I Ideal Effort Force (lb)	E_A Actual Effort Force (lb)	IMA Ideal Mechanical Advantage	AMA Actual Mechanical Advantage	η Efficiency (%)
A	20.00		8.20	22.65	20.00	6.82	9.89			
B	100.00	24.00			20.00			2.90	2.20	
C			96.00	132.54			88.91		1.10	.66
D	50.00		96.00		30.00	25.00		2.00	1.00	

FIGURE 3-4 *A table of inclined planes.*

Inclined Plane A:

Slope length: ______________________ IMA: ______________________

AMA: ______________________ η: ______________________

Inclined Plane B:

Height: ______________________ Width: ______________________

E_I: ______________________ E_A: ______________________

η: ______________________

Inclined Plane C:

Load: ______________________ Slope length: ______________________

Slope angle: ______________________ E_I: ______________________

IMA: ______________________

Inclined Plane D:

Slope length: ______________________ Width: ______________________

E_A: ______________________ η: ______________________

Problem 3.3 A hydraulic-powered log splitter, like the one shown in Figure 3-5A, is used to cut large logs in half. A wedge, located on one end of the device, provides the cutting action. The shape of the wedge is an isosceles triangle that is 6 in. thick by 12 in. wide. An *actual* effort force of 300 lb is required to split a log. The cutting action of the wedge is 80% efficient. Use this information to answer the following questions. Show your math work in the spaces provided.

(A)

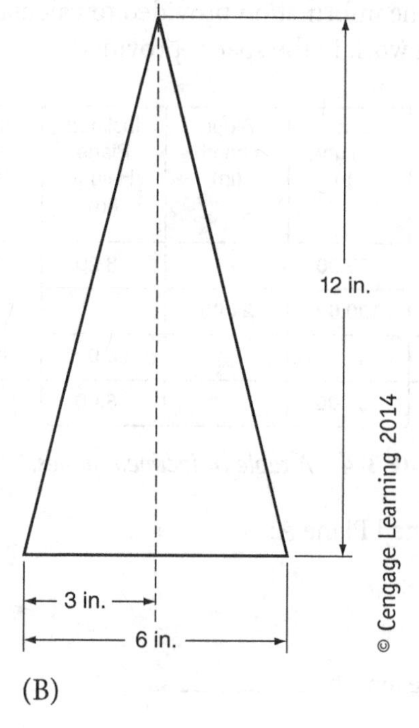

(B)

FIGURE 3-5 *(A) Log splitter and (B) wedge profile.*

1. What is the slope length (s) on either side of the wedge?

$s =$ ______________________

2. What is the ideal mechanical advantage (IMA) of the wedge?

IMA = ______________________

3. What is the actual mechanical advantage (AMA) of the wedge?

AMA = ______________________

4. How much resistance does the log provide against the cutting action of the wedge?

Load = ______________________

5. If friction is not a consideration, what would the ideal effort force (E_I) need to be to split the log?

E_I = ______________________

Problem 3.4 Levers X, Y, and Z in Figure 3-6 represent three different lever systems. Identify which of these three lever systems is described in each of the statements below. Write the answers and justification for each selection in the spaces provided.

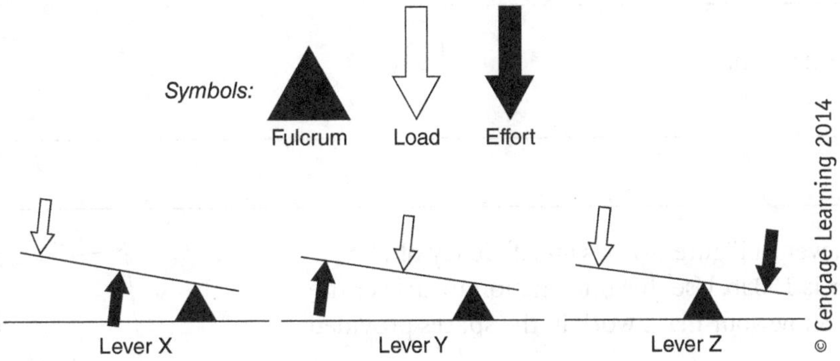

Symbols: Fulcrum Load Effort

Lever X Lever Y Lever Z

© Cengage Learning 2014

FIGURE 3-6 *Lever systems X, Y, and Z.*

1. This system represents a class of lever that always has a mechanical *disadvantage.*

____

Explain your selection.

____

____

2. This represents a class of lever that always has a mechanical *advantage.*

____

Explain your selection.

____

____

3. This is an example of a first-class lever.

____

Explain your selection.

____

____

4. This is an example of a second-class lever.

____

Explain your selection.

____

____

5. This is an example of a third-class lever.

____

Explain your selection.

____

____

Problem 3.5 The lever in Figure 3-7 has an efficiency of 75%. The load on the lever is 150 lb. Use this information to answer the following questions. Show your math work in the spaces provided.

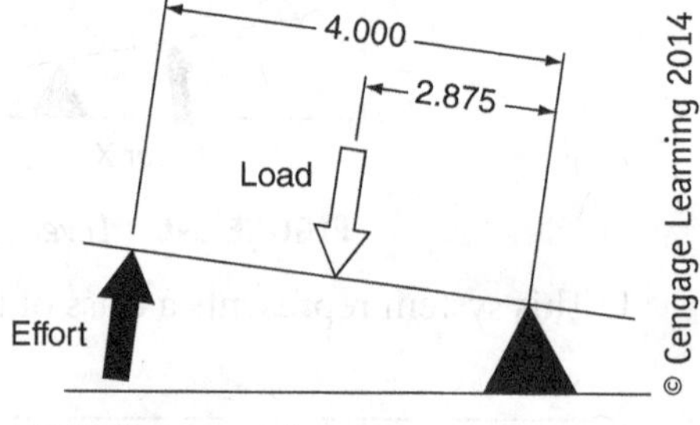

FIGURE 3-7 *A lever.*

1. Figure 3-7 is an example of what class of lever?

____

2. What is the ideal mechanical advantage (IMA) of the lever?

IMA = ______________________________

3. What is the actual mechanical advantage (AMA) of this system?

AMA = ______________________________

4. What is the *ideal effort force* (E_I) that would be needed to move the lever in the absence of friction?

E_I = ______________________________

5. What is the actual effort force (E_A) that must be applied to the lever to move the load?

E_A = ______________________________

6. How could you alter this machine to increase its ideal mechanical advantage (IMA)?

____

____

Problem 3.6 In Figure 3-8, a 40-lb bucket of water is being raised from a deep well using a winch system that is 85% efficient. The winch axle diameter is 4 in., and the crank handle is located 12 in. from the center axis of the axle. Use this information to answer the following questions. Show your math work in the spaces provided.

© iStockphoto/alexsol

FIGURE 3-8 *A wheel and axle system.*

1. What is the ideal mechanical advantage (IMA) of the wheel and axle?

IMA = ______________________________

2. What is the actual mechanical advantage (AMA) of the wheel and axle?

AMA = ______________________________

3. Under *ideal* conditions, how much effort must be applied to the crank handle to raise the bucket of water?

E_I = ______________________________

4. What is the *actual* amount of effort force (E_A) that must be applied to the crank handle to raise the bucket of water?

E_A = ______________________________

Problem 3.7 The pulley system in Figure 3-9 is 70% efficient. The 160-lb load must be moved a vertical distance of 72 in. Use this information to answer the following questions. Show your math work in the spaces provided.

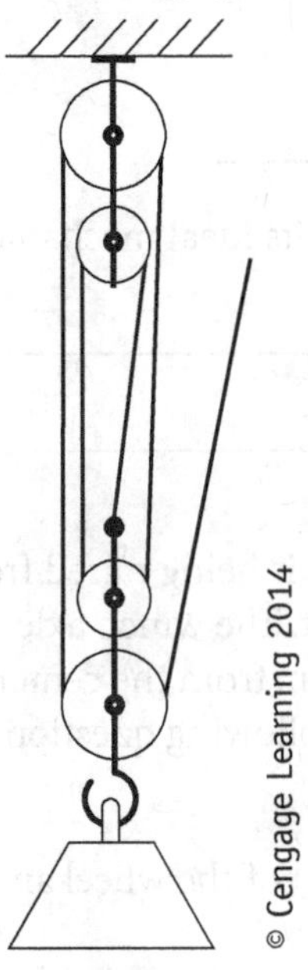

FIGURE 3-9 *A pulley system.*

1. What is the ideal mechanical advantage (IMA) of this pulley system?

   IMA = ______________________________

2. What is the actual mechanical advantage (AMA) of this pulley system?

   AMA = ______________________________

3. What is the length of the effort cable (d_E) that must pass through the user's hands to raise the load to the desired height?

   d_E = ______________________________

4. What is the ideal effort force (E_I) that would be needed to move the load using the pulley system?

   E_I = ______________________________

5. What is the actual effort force (E_A) that would be needed to move the load using the pulley system?

   E_A = ______________________________

Problem 3.8 The vise in Figure 3-10 contains a 3/8-16 UNC-threaded rod. A 500-lb holding force can be generated by turning the 4-in.-long handle. The actual mechanical advantage (AMA) of the machine is 250. Use this information to answer the following questions. Show your math work in the spaces provided.

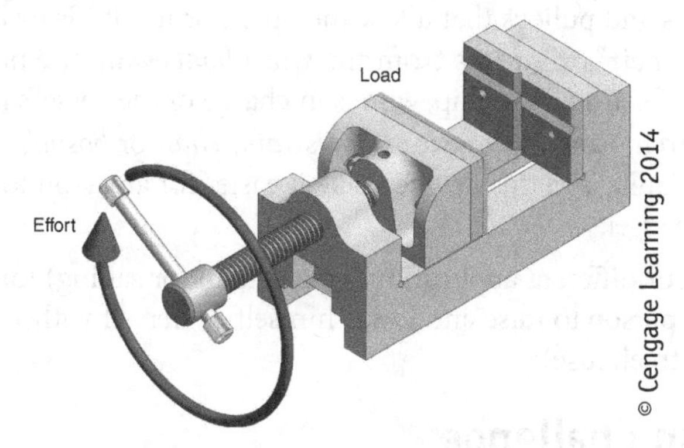

FIGURE 3-10 *A screw-actuated vise.*

1. What is the pitch of the screw thread? *See Equation 3-14 in your textbook.*

   p = ______________________

2. What is the ideal mechanical advantage (IMA) of the screw?

   IMA = ______________________

3. What is the efficiency (η) of the screw system?

   η = ______________________

4. What is the ideal effort force (E_I) that must be applied to the vise handle to hold a load?

   E_I = ______________________

5. What is the actual effort force (E_A) that must be applied to the vise handle to hold a load?

   E_A = ______________________

Explore Your World

A boatswain's or bosun's chair (pronounced *bōsŭn*) is a device that is typically used to work aloft on the rigging of a sailing vessel. The apparatus generally consists of a seat or harness in which an individual sits, and it is connected to a series of ropes and pulleys that allow the operator to lift his or her own body weight without assistance. The name "bosun's chair" originates from the word *boatswain,* the name for the petty officer on a merchant ship or the warrant officer on a warship, who is in charge of the vessel's rigging and equipment.

- Conduct an Internet search using the keyword *boatswain's chair* or *bosun's chair*, to find more information on how the pulleys in this device make it easier for a person to hoist himself or herself up the mast of a sailing vessel.
- Create a list of at least four different applications (other than for sailing) for the bosun's chair where it might be desirable for a person to raise and lower himself or herself without the assistance of another (e.g., a lift elevator for a treehouse).

Exercise 3.3 Design Challenge

Objective

At the conclusion of this design challenge, you will be able to do the following:

1. Transfer what you have learned about the function of a bosun's chair to designing a unique, self-powered lift device that uses a pulley system.
2. Communicate design ideas through annotated sketches.
3. Demonstrate the effectiveness of a device by calculating its mechanical advantage and ideal effort requirement.

Procedure

Complete the research on the bosun's chair as outlined in the previous *Explore Your World* section. Generate a list of four unique applications for a bosun's chair-style device. *Keep in mind that an alternative design may allow the operator to be standing or lying down rather than sitting.*

Design Brief	
Problem Statement:	There are many circumstances where it would be advantageous to have a human-powered lift device that can be operated by the person who is being transported.
Design Statement:	Design a unique lift device to be used for a specific application, wherein the operator can be transported using a pulley system to lift his or her own body weight.
Constraints:	1. The maximum ideal effort must be no more than 20 lb. 2. The apparatus must use a pulley system, a safety restraining device, and a support for the operator/rider (chair, sling, platform, gurney, harness, etc.). 3. The pulley system must be designed to accommodate the weight of the operator, as well as the weight of the support device.
Deliverables:	• A pictorial sketch of the entire lift apparatus, including the pulley system and cables, restraining device, and support mechanism. The sketch must be annotated with rough dimensions, notes about how the apparatus works, and the height that it must travel. *See Figure 3-43 on page 96 in your textbook for an example of an annotated sketch.* • A detailed sketch of the pulley system. The sketch must be annotated with cable lengths, the system's load capacity, and calculations for the mechanical advantage of the system and for the ideal effort needed to lift the intended load.

CHAPTER 4
Mechanisms

Before You Begin

Think about these questions as you study the concepts in this chapter.

- What are the different types of linkage systems, and what are their applications?
- What is the purpose of a cam and follower, and what kinds of products contain them?
- What is the difference between a simple and compound gear train?
- How do you calculate the gear ratio, speed ratio, and torque ratio of a simple and compound gear train?
- What components make up a chain drive mechanism?
- What factors must be considered when an engineer designs a chain drive mechanism?

Procedure

Read the section on linkages in Chapter 4, "Mechanisms," in your textbook.

Explore Your World

Locate at least one device in your home that uses a linkage system (for example, see Figure 4-1a, b, and c on page 102 in the *Principles of Engineering* textbook). Create a pictorial sketch (refer to Figure 2-12 on page 45 of your textbook) or multiview sketch (refer to Figure 2-13 on page 46 of your textbook) of the device, and label any rockers, cranks, connecting bars, sliders, or pin joints. Explain how the linkage system operates to control the motion of your device.

Cam and Follower

Exercise 4.1 Calculating Cam Displacement

Objective

At the conclusion of this exercise, you will be able to do the following:

1. Identify the base circle of a cam.
2. Determine the shortest and longest radial distance of a cam.
3. Create a cam displacement diagram to identify the rise, drop, and dwell cycles of a specific cam.

Procedure

Read the section on the cam and follower in Chapter 4 in your textbook.

Materials

- Compass (drafting)
- Standard inch rule with an accuracy of 1/16 inch (see Figure 14-4 on page 449 of your textbook)

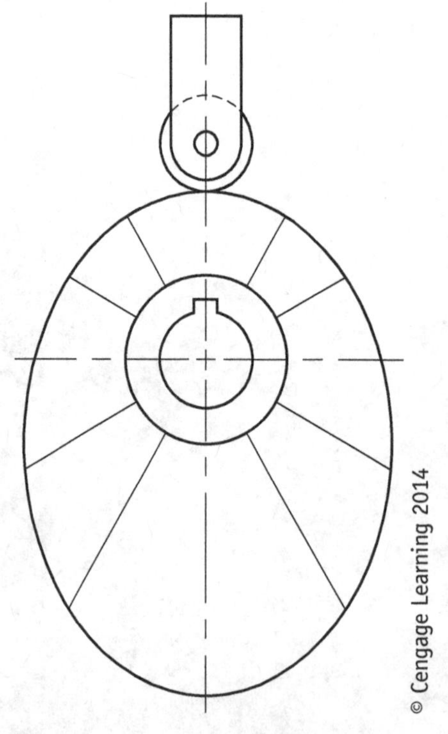

FIGURE 4-1 *A cam and follower schematic.*

Complete the following exercises for the cam and follower system in Figure 4-1.

1. Carefully draw the base circle on Figure 4-1 using a compass.
2. Label the radial divisions on Figure 4-1 starting at 0°, where the follower is shown in contact with the cam profile, and continuing clockwise in 30° increments. (For example, see Figure 4-12a on page 108 of your textbook.)
3. Measure the shortest radial distance for this cam and record the measurement and angle in the space provided.

   Shortest radial distance ___________ in. at ____°

4. Measure the longest radial distance for this cam and record the measurement and angle in the space provided.

   Longest radial distance ___________ inch at ____°

5. Calculate the total follower displacement for this cam. Show your math work in the space below.

   Total follower displacement ______________

6. Using the chart provided in Figure 4-2, draw the displacement diagram for the cam and follower in Figure 4-1. (Assume that the follower is currently at 0° and that the cam rotates counterclockwise.) Label the rise, drop, and dwell periods on the cam displacement diagram. For an example, see Figure 4-12b on page 108 of your textbook.

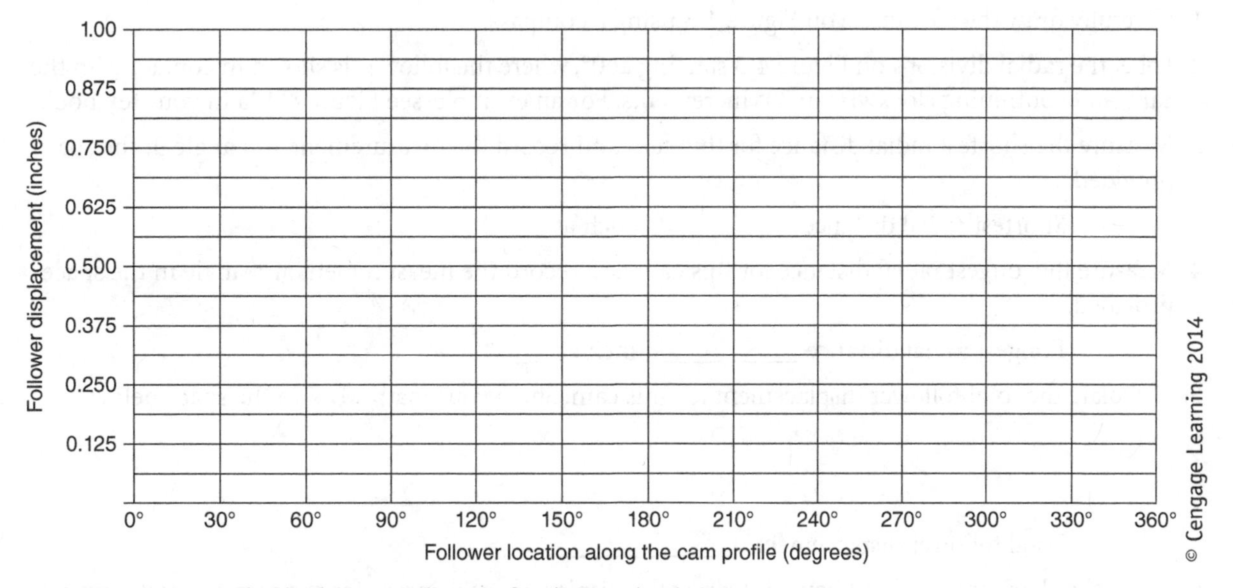

FIGURE 4-2 *Draw the displacement diagram for the cam in Figure 4-1.*

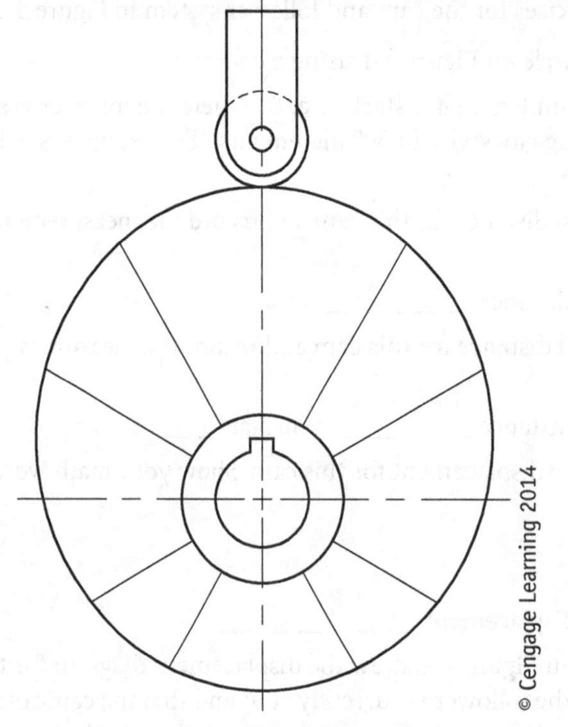

FIGURE 4-3 *A different cam and follower schematic.*

Complete the following exercises for the cam and follower system in Figure 4-3.

1. Carefully draw the base circle on Figure 4-3 using a compass.
2. Label the radial divisions on Figure 4-3 starting at 0°, where the follower is shown in contact with the cam, and continuing clockwise in 30° increments. For an example, see Figure 4-12a in your textbook.
3. Measure the shortest radial distance for this cam and record the measurement and angle in the space provided.

   Shortest radial distance ___________ inch at ____°

4. Measure the longest radial distance for this cam and record the measurement and angle in the space provided.

   Longest radial distance ___________ inch at ____°

5. Calculate the total follower displacement for this cam. Show your math work in the space below.

   Total follower displacement ___________

6. Using the chart provided in Figure 4-4, draw the displacement diagram for the cam and follower in Figure 4-3. (Assume that the follower is currently at 0° and that the cam rotates counterclockwise.) Label the rise, drop, and dwell periods on the cam displacement diagram. For an example, see Figure 4-12b in your textbook.

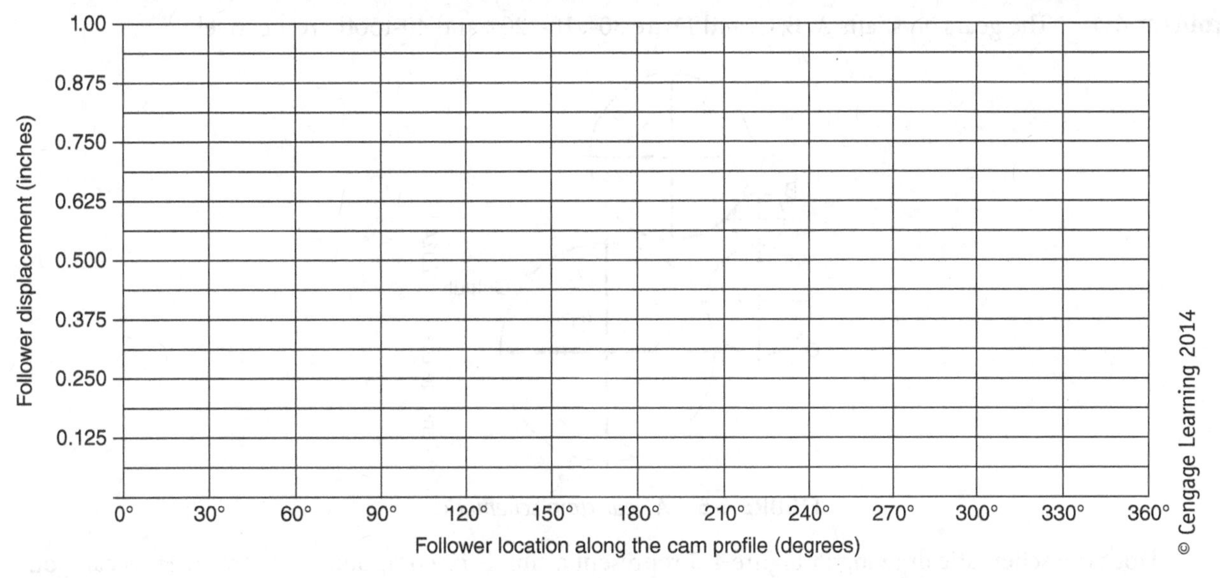

FIGURE 4-4 *Draw the displacement diagram for the cam in Figure 4-3.*

Gear Trains

Exercise 4.2 Calculating Gear Train Speed and Torque Ratios

Objective

At the conclusion of this exercise, you will be able to do the following:

1. Interpret a schematic drawing of a gear train.
2. Calculate the gear ratio for a simple and compound gear train.
3. Interpret a gear ratio as indicating either a speed-increasing or speed-reducing gear train.
4. Identify the relationships between the input and output of a gear train.
5. Calculate the rotational speed of an output gear shaft.
6. Calculate the torque of an output gear shaft.

Procedure

Read the section on gears in Chapter 4 in your textbook. Complete the following problems.

TIP SHEET

The gear ratio (GR), speed ratio (SR), and torque ratio (TR) of a gear train should be expressed in a ratio format, not as a fraction. For example, a speed-increasing gear train would be expressed as 1:*n* and a speed-reducing gear train would be expressed as *n*:1 (in each case, *n* represents a number greater than 1). When you calculate the gear ratio of a compound gear train, you should never return a value for *n* that is less than 1. An erroneous ratio, such as 1:0.75, may indicate that you have inadvertently used the reciprocal value of one of the mating gear sets in your calculations.

Problem 4.1 The gears on shafts A, B, C, and D are 30-, 10-, 20-, and 40-tooth, respectively.

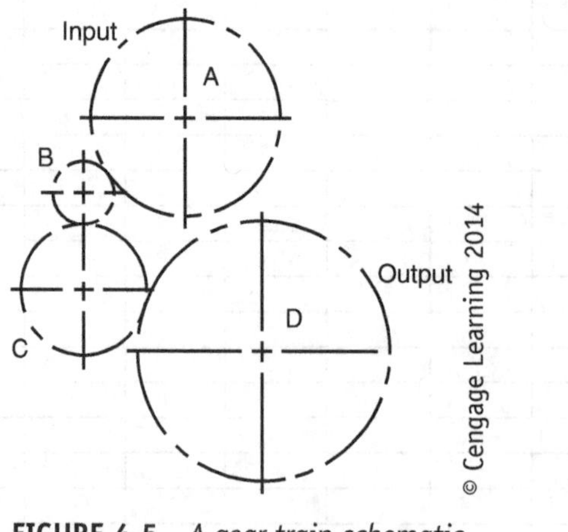

FIGURE 4-5 *A gear train schematic.*

1. Does the schematic drawing in Figure 4-5 represent a simple or compound gear train? How can you tell?

2. What is the gear ratio for this gear train? Does it represent a speed increase or decrease?

3. If the gear on shaft A is turning in a clockwise direction, in what direction will the gear on shaft D rotate?

4. Label the direction of rotation of each of the gears in the gear train with a curved arrow (for example, ↻ or ↺).

Problem 4.2 The gears on shafts A, Z, and D are 30-, 40/10-, and 40-tooth, respectively.

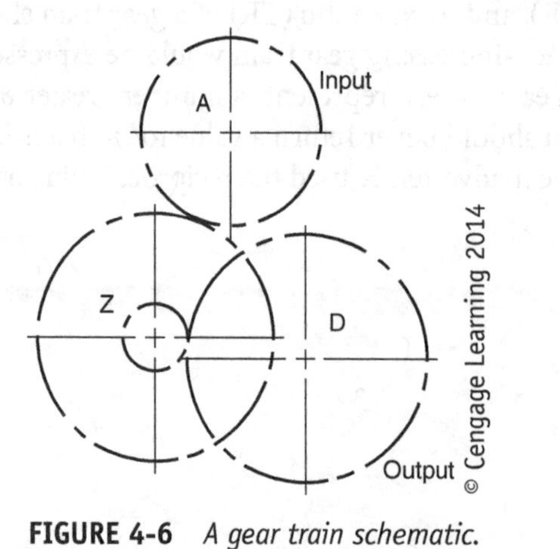

FIGURE 4-6 *A gear train schematic.*

1. Does the schematic drawing in Figure 4-6 represent a simple or compound gear train? How can you tell?

2. What is the gear ratio for this gear train? Does it represent a speed increase or decrease?

3. If the gear on shaft A is turning in a clockwise direction, in what direction will the gear on shaft D rotate?

4. Label the direction of rotation of each of the gears in the gear train with a curved arrow (for example, ↺ or ↻).

Problem 4.3 An electric motor is connected to the input shaft on the compound gear train shown in Figure 4-7. The motor spins at 5,000 RPM and generates 20 lbf·ft of torque. The gears on shafts A, B_1, B_2, C_1, C_2, and D are 10-, 40-, 20-, 40-, 30-, and 20-tooth, respectively.

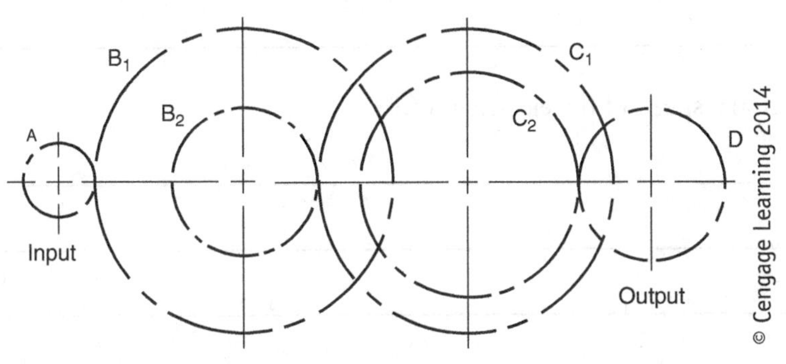

FIGURE 4-7 *A different gear train schematic.*

1. What is the gear ratio for this gear train? Does it represent a speed increase or decrease?

2. What is the rotational speed of the output gear shaft?

3. What is the torque value on the output gear shaft?

Problem 4.4 An electric motor is connected to the input shaft on the compound gear train shown in Figure 4-8. The output gear spins at 140.58 RPM and generates 106.7 lbf·ft of torque. The gears on shafts A, B_1, B_2, C_1, C_2, and D are 10-, 40-, 30-, 40-, 20-, and 40-tooth, respectively.

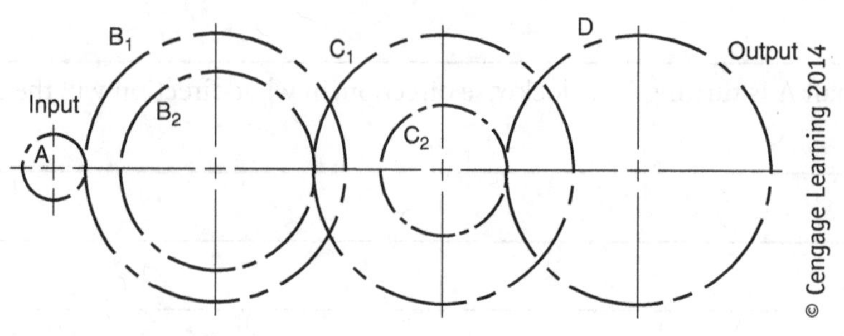

FIGURE 4-8 *Another gear train schematic.*

1. What is the gear ratio for this gear train? Does it represent a speed increase or decrease?

2. What is the rotational speed of the electric motor?

3. How much torque is the electric motor exerting on the input shaft?

Sprockets and Chain Drives

Exercise 4.3 Calculating Speed and Torque Ratios for Chain Drive Mechanisms

Objective

At the conclusion of this exercise, you will be able to do the following:

1. Calculate the chain drive ratio for a chain drive system.
2. Calculate the rotational speed of the output sprocket in a chain drive system.
3. Calculate the torque value of the input sprocket in a chain drive system.

Procedure

Read the section on sprockets and chain drives in Chapter 4 in your textbook. Complete the problem below.

© iStockphoto/mipan

FIGURE 4-9 *A bicycle chain and sprocket mechanism.*

Problem 4.5 A BMX bicycle uses a pedal-operated, 40-tooth sprocket that is connected via a roller chain to a rear-axle, 15-tooth sprocket. The pedal-driven sprocket rotates at 63 RPM, and the output sprocket has 1,000 lbf·ft of torque. Use this information to answer the following questions:

1. What is the chain drive ratio for the chain drive system shown in Figure 4-9?

____

____

____

____

2. What is the rotational speed of the output sprocket?

____

____

____

____

3. What is the torque value on the input sprocket?

Explore Your World

To see an example of how an oscillating lawn sprinkler uses both gears and cams to control the speed and duration of the device's back-and-forth motion, explore the following Web site: http://home.howstuffworks.com/sprinkler1.htm, or conduct an Internet search using the keywords *oscillating sprinkler, cams.*

CHAPTER 5
Energy

Before You Begin

Think about these questions as you study the concepts in this chapter.

- What are the types of energy?
- How is energy transferred?
- What are some examples of energy transformation?
- What are the units for work and energy?
- Is all the energy that is put into a system transferred to useful work?

Exercise 5.1 Energy Transformation Match-Up

Objective

The purpose of this activity is to match the correct energy transformation to the item or event.

Materials

- Scissors
- Energy Transformation Match-Up Puzzle

Procedure

Cut out the Energy Transformation Match-up Puzzle pieces and mix them up on your desk. Your task is to match up the item with the appropriate energy transformation. The challenge is that there will be unmatched items on the perimeter of your cube. See the example shown in Figure 5-1.

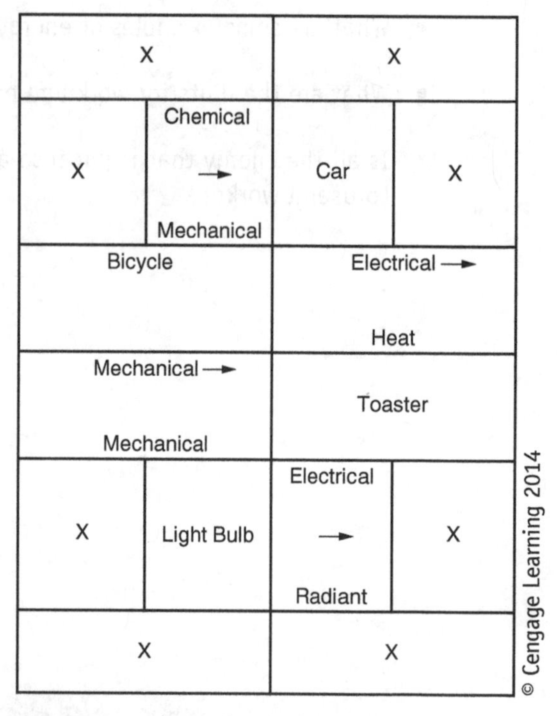

FIGURE 5-1 *Sample match-up puzzle.*

"Chemical → Mechanical" matches up with "Car," "Bicycle" matches up with "Mechanical → Mechanical," "Toaster" matches up with "Electrical → Heat," and "Light Bulb" matches up with "Electrical → Radiant."

When you finish, have your teacher check your puzzle and initial in the space provided:

Teacher's Initials:______________________________ Date:______________________

Energy Transformations Match-Up Puzzle

<table>
<tr>
<td>Electrical →
Radiant
Chemical → Smoke
Radiant Detector
X-Ray</td>
<td>Radiometer
Thermal
Light Bulb →
Sound
Motion →
Sound</td>
<td>Nuclear
Electrical
Hand
Radio
Warmer
Electrical →
Motion</td>
<td>Electrical →
Heat
Electrical
Battery →
Radiant
Nuclear Power
Plant</td>
</tr>
<tr>
<td>Glow Stick
Coal
Electrical →
Power
Sound
Plant
Microwave</td>
<td>Bicycle
Nuclear
PE
Tree
→
Sound
Motion →
Light</td>
<td>Windmill
Rubbing
Sun Hands
Together
Rocket</td>
<td>Analog Film
Chemical →
Piano
Radiant
Electrical →
Motion</td>
</tr>
<tr>
<td>Acoustic Guitar
Chemical →
TV
Electrical
Motion →
Motion</td>
<td>Chemical →
Mechanical
Chemical
Magnifying
→
Glass
Electrical
Radiant →
Electrical</td>
<td>Clothes Iron
Geiger Fire
Counter Cracker
Radiant →
Motion</td>
<td>Fan
Candle Car
Washing Machine</td>
</tr>
<tr>
<td>Wind-up Toy
Chemical Electrical
→ →
Sound Motion
Radiant →
Sound</td>
<td>Hand-powered
Flashlight
Electrical
Tanning
→
Bed
Radiant
Chemical →
Radiant</td>
<td>Satellite Dish
Motion → Radiant →
Thermal Chemical
Curling
Iron</td>
<td>Photovoltaic Cell
Chemical
Mixer →
Thermal
Radiant →
Chemical</td>
</tr>
</table>

Exercise 5.2 Light Bulb or Heat Bulb?

Objective

The purpose of this lab is to determine which type of light bulb is more efficient at converting electrical energy to radiant energy.

Materials

- Compact fluorescent light bulb (CFL)
- Incandescent light bulb (IL)
- Two (2) lamp bases
- Fahrenheit thermometer
- Clock or timer
- Two (2) empty pop cans covered with black electrical tape or black paper, or painted black
- Water
- Graduated cylinder

Procedure

1. Insert a CFL bulb in one lamp base, plug the lamp in, and turn it on.
2. Insert an IL bulb in one lamp base, plug the lamp in, and turn it on.
3. Place 20 ml of water in each pop can.
4. Record initial (time = 0 min) water temperature in °F in the data table below.
5. Place one can 20 cm away from the CFL bulb.
6. Place one can 20 cm away from the IL bulb.
7. Check and record the water temperature at 5-minute intervals and record the resulting data in Table 5-1.

The Celsius or Kelvin scale is primarily used in science and engineering. However, because water has such a high specific heat and is very resistant to temperature changes, it can be very difficult to measure a change of water temperature in degrees Celsius or Kelvin. Because degrees Fahrenheit is a smaller unit, you will be using that to record water temperature, and then converting to degrees Celsius.

	0 min	5 min	10 min	15 min	20 min	25 min	Change in Temp (°F) ($\Delta T = T_f - T_i$)
Water temperature in CFL can (°F)							
Water temperature in IL can (°F)							

TABLE 5-1 *Water Temperature Comparison.*

Analysis

1. Convert the change in temperature, ΔT, for the water temperature for the CFL can and the water temperature for the IL can from °F to °C.

2. Calculate the heat energy used to heat the water to each temperature. The specific heat of water is 4180 J/(kg · °C).

3. Construct a line graph showing the temperature change as a function of time using either graph paper or Microsoft Excel. Be sure to include a title and axis labels.

Conclusion

1. How was energy transferred from the bulb to the water?

2. Which light bulb was more efficient at converting electrical energy to radiant energy?

3. Using dimensional analysis, show that the units in the formula $Q = mc\Delta T$ work out.

Exercise 5.3 Follow the Bouncing Ball Energy Transformations

Objective

The purpose of this activity is to examine energy transformations in a system and calculate the efficiency of a system.

Materials

- Meter stick
- Ball that can bounce

Pre-Lab Questions

1. When dropped from a height of 1 meter, how high do you predict the bouncing ball will rebound?

____

2. What is the law of conservation of energy?

Procedure

Obtain a ball that can bounce and a meter stick from your teacher. Have one group member hold the meter stick and drop the ball from a height of 1 meter. Have another group member spot the ball's rebound height. Have another student record that height in meters. Do at least three trials.

	Rebound Height (m)	Efficiency
Trial 1		
Trial 2		
Trial 3		
Average Values		

TABLE 5-2 *Bouncing Ball Energy Transformations.*

Next, calculate the efficiency of the ball by using the following equation:

$$Eff = \frac{h_{rebound}}{h_{initial}}$$

and record in the data table.

Calculate the average values for rebound height and efficiency and record in the spaces provided in Table 5-2.

Next, vary the drop height for each of the five trials. Record the rebound height, calculate the efficiency, and record the results in Table 5-3.

	Drop Height (m)	Rebound Height (m)	Efficiency
Trial 1			
Trial 2			
Trial 3			
Trial 4			
Trial 5			
Average Values			

TABLE 5-3 *Varying Drop Heights.*

Conclusion

1. Describe *all* the energy transformations that the bouncing ball went through from the initial drop, to the impact on the table, to the rebound. This should read like a flowchart.

2. What are the units of efficiency measured in?

3. Was energy conserved during this experiment?

4. (a) Why did the ball not rebound as high as the original height?

(b) The lost energy was transferred into what two types of energy?

5. How does the drop height affect the efficiency of the bouncing ball?

CHAPTER 6
Electrical Systems

Before You Begin

Think about these questions as you study the concepts in this chapter.

- What types of materials make good conductors of electricity?
- What types of materials are good electrical insulators?
- How can resistors be used to manipulate the voltage across and current through components in a circuit?
- What are the units for voltage, current, resistance, and power?
- Which types of appliances use the most power?

Exercise 6.1 Conductors and Insulators

Objective

The purpose of this lab is to determine which materials are conductors and insulators.

Materials

- Multimeter
- Battery pack with 9 V total voltage (may have one 9-volt battery, or six 1.5-volt batteries connected in a battery pack).
- Christmas light bulb or other small light bulb
- Alligator clips
- Penny
- Plastic bingo chip
- Aluminum foil
- Wood toothpick or skewer
- Nickel
- Cardboard or heavy paper

Pre-Lab Questions

Rank the following items in Table 6-1 from 1–6 according to which items are the best conductors. A "1" indicates the best conductor, while a "6" indicates the best resistor.

	Plastic bingo chip
	Penny
	Aluminum foil
	Wood toothpick or skewer
	Nickel
	Cardboard or heavy paper

TABLE 6-1 *Conductor Rankings (Pre-Lab).*

Procedure

Use a multimeter to obtain the value of each object's resistance. Record each of the object's resistances in Table 6-2, and then rank them according to which is the best conductor. How do your answers compare to your initial ideas?

Item	Resistance (Ω)
Plastic bingo chip	
Penny	
Aluminum foil	
Wood toothpick or skewer	
Nickel	
Cardboard or heavy paper	

TABLE 6-2 *Conductor Rankings (Lab).*

Connect the battery and lamp with alligator clips in a series circuit, as shown in Figure 6-1. Note the brightness of the lamp.

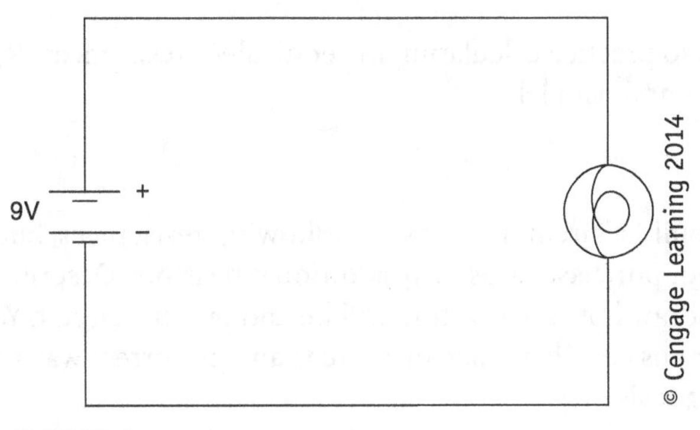

FIGURE 6-1

Connect each of the above items in series with the lamp. If the light is lit, record if it appears bright or dim in Table 6-3.

Item	Is the light lit?	If the light is lit, is it bright or dim?
Plastic bingo chip		
Penny		
Aluminum foil		
Wood toothpick or skewer		
Nickel		
Cardboard or heavy paper		

TABLE 6-3 *Level of Resistance.*

Conclusion

1. *In general,* which types of items made the best conductors?

2. *In general,* which types of items made the best insulators?

3. Which item was the best conductor?

4. Which item was the best insulator?

Exercise 6.2 Resistor Challenge

Objective

The purpose of this activity is to practice calculating the equivalent resistances, R_{eq}, for resistors in series, parallel, and a combination of series and parallel.

Procedure

Your employer wants you to build a circuit that has the following resistances, but you have only a few resistors to accomplish this. You may not purchase or use any additional resistors. Describe how you will do this for each design requirement. You may draw how the resistors will be laid out in a circuit. You must show the calculations that you used to arrive at your answer. There may be more than one correct way to accomplish this.

You are given the following resistors:

- 2 10-kΩ resistors
- 1 5-kΩ resistors
- 4 1-kΩ resistors
- 3 500-Ω resistors

Problem 6.1 Equivalent resistance must be equal to 25 kΩ.

Problem 6.2 Equivalent resistance must be equal to 250 Ω.

Problem 6.3 Equivalent resistance must be equal to 750 Ω.

Problem 6.4 Equivalent resistance must be equal to 400 Ω.

Exercise 6.3 More Power Lab

Objective

The purpose of this lab is to determine which types of appliances use the most power.

Materials

- Kil-A-Watt Meter(s)
- Various small appliances, such as the following:
  - o Overhead projector
  - o Fans
  - o Incandescent light bulb (IL) with lamp base
  - o Compact fluorescent light bulb (CFL) with lamp base
  - o Hair dryer with multiple settings
  - o Small radio
  - o Night-light
  - o Vacuum cleaner

Procedure

Record in Table 6-4 the values from the Kil-A-Watt meter for Voltage, Current, Resistance, and Power for each appliance tested. If more than one Kil-A-Watt meter is available, walk around the room to each station. Alternatively, this may be done as a teacher demo if only one Kil-A-Watt meter is available.

Appliance	Voltage (V)	Current (Amps)	Measured Power (W)	Calculated Power (W)	Calculated Resistance (Ω)

TABLE 6-4 *Appliance Readings.*

Conclusion

1. Which appliances used the most energy? Do they have anything in common?

2. Which appliances used the least energy? Do they have anything in common?

3. Some appliances may have a different power reading than what you calculated. Why do you think this may happen?

CHAPTER 7
Fluid Power Systems

Before You Begin

Think about these questions as you study the concepts in this chapter.

- What are some advantages and disadvantages of using pneumatic and hydraulic systems?
- What are the English and metric units for pressure?
- What are some examples of energy transformation?
- What are the units for work and energy?
- Is all the energy that is put into a system transferred to useful work?

Exercise 7.1 High Pressure

The purposes of this exercise are to determine the initial velocity of Diet Coke leaving a bottle and the internal pressure of the liquid after Mentos candies have been added.

Objective

The purpose of this activity is to calculate the internal pressure of a Diet Coke and Mentos geyser based on the height.

Materials

- 2-liter bottle of Diet Coke
- 1 pkg. of Mentos Mint candies
- A long metric tape measurer or several meter sticks
- Splash goggles
- Rain poncho (optional)

Procedure

This may be done as a small group activity or as a teacher demo.

1. It is best to perform this experiment/demo outdoors. However, if not, place a large plastic tarp underneath the bottle. Make sure to have plenty of space available—this will be messy!
2. Record the mass of the full Diet Coke bottle in kilograms: ______________kg.
3. Place the bottle on a level surface.
4. Remove the cap from the bottle.
5. The goal is to get all the Mentos into the bottle as quickly as possible. Putting the Mentos in a test tube with a large enough diameter or a similar type of tube may be helpful.
6. Warn participants to place all the Mentos in the bottle quickly and run back to a safe distance.
7. Record the height of the geyser in meters: ___________m. The height may go from 0.5 meters to up to 10 meters! Be prepared to get accurate measurements for this height range.
8. Empty and rinse out the bottle. Record the mass of the empty bottle in kilograms: ___________ kg.
9. Calculate the mass of the liquid and record it in the space below. The mass of the Mentos will be considered negligible.______________kg.

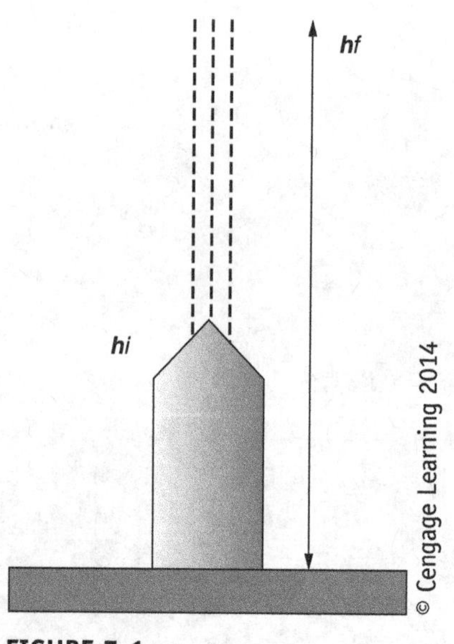

FIGURE 7-1

Analysis

Determining the Initial Velocity of the Diet Coke The initial kinetic and potential energy must be equal to the final kinetic and potential energy given by the equation below:

$$PE_i + KE_i = PE_f + KE_f$$

The equation in its expanded form is below:

$$mgh_i + \frac{1}{2}mv_i^2 = mgh_f + \frac{1}{2}mv_f^2$$

Because $h_i = 0$ m, the initial potential energy value goes to zero, and because the final velocity at the final height of the geyser is zero, the final kinetic energy term also goes to zero:

$$\cancel{mgh_i} + \frac{1}{2}mv_i^2 = mgh_f + \cancel{\frac{1}{2}mv_f^2}$$

Solve the above equation for the initial velocity,

$$v_i =$$

Using $g = 9.8\ \text{m/s}^2$, and your measurements from above, solve for the initial velocity:

$$v_i =$$

Determining Pressure To determine the pressure of the Diet Coke after the Mentos have been dropped into the bottle, the following form of Bernoulli's law will be used:

$$P = \frac{1}{2}dv_i^2$$

where d is the density of the Diet Coke.

To find the density (in kg/L), take the mass of the Diet Coke and then divide it by the volume in *liters*:

$$d = \_\_\_\_\_\_\_\_\_\_\_\_$$

Then substitute the equation for density and the value for the initial velocity into the above pressure equation to calculate the pressure caused by the reaction of the Mentos with the Diet Coke:

$$P = \_\_\_\_\_\_\_\_\_\_\_\_$$

Exercise 7.2 That Sinking Feeling

Objective

The purpose of this exercise is to determine the downward force needed to create neutral buoyancy on an object.

Materials

- A small rectangular aquarium or other clear tank
- Electronic scale or triple beam balance
- Water
- Masses of various sizes (may be laboratory masses, metal washers, metal hexagonal nuts, etc. . . .)

Possible materials for this lab may include:

- Small block of wood
- Ice cube
- Ping-pong ball

Procedure

1. Carefully fill an aquarium to about 75% capacity.
2. Record the following information.
3. Place the object in the water so that it floats. Roughly what percentage of the object is below the water? Draw and label a diagram of the object floating:

4. Attach various masses to the object until it reaches neutral buoyancy.
5. Once the object reaches neutral buoyancy for a few moments, take the object out, remove the masses, and record the total mass of your object. ______________kg.
6. Convert this mass to a force by multiplying it by 9.8 m/s2. Record this force here: ___________________N.

Conclusion

1. Compare your results with your classmates. Which object required the *most* force to obtain neutral buoyancy? Which object required the *least* force to obtain neutral buoyancy?
2. Challenge question: Explain how a ship that is just slightly taller than the underpass of a bridge can use Bernoulli's law to go under that bridge safely.

Exercise 7.3 Pneumatic and Hydraulic Products

Objective

The purpose of this exercise is to examine products that use pneumatic and hydraulic systems.

Materials

- A computer with Internet access
- Word processing program (optional)

Procedure

Pneumatic, Hydraulic, or Both?—Part 1 Identify the components in Table 7-1 as *pneumatic* (P), *hydraulic* (H), or *both* (B).

1.		Filter
2.		Pilot line
3.		Check valve
4.		Pump
5.		T-connector
6.		Air compressor
7.		Piston
8.		Shutoff valve
9.		Directional control valve
10.		Valve
11.		Variable displacement pump
12.		Actuator
13.		Solenoid
14.		Double-acting cylinder
15.		Lubricator
16.		Receiver tank
17.		Flow control valve
18.		Fixed-displacement pump
19.		Prime mover
20.		Working line
21.		Lubricator
22.		Reservoir
23.		Single-acting cylinder
24.		Pressure regulator

TABLE 7-1 *Component Identification.*

Pneumatic, Hydraulic, or Both?—Part 1

1. Select one pneumatic and one hydraulic product that you would like to investigate.
2. Use the Internet to find out how these products work.
3. Describe how each device works.
4. Be sure to highlight the following components, typically found in each device. (Keep in mind, however, that your device may not have all of them.)
    a. Working line
    b. Pilot line
    c. Valve
    d. T-connector
    e. Filter
    f. Actuator
    g. Single-acting cylinder
    h. Piston
    i. Double-acting cylinder
    j. Shutoff valve
    k. Check valve
    l. Shuttle valve
    m. Flow-control valve
    n. Reservoir
    o. Pump
    p. Fixed-displacement pump
    q. Variable-displacement pump
    r. Prime mover
    s. Directional control valve
    t. Air compressor
    u. Pressure regulator
    v. Lubricator
    w. Receiver tank
    x. Solenoid
5. Cite all sources using APA format

CHAPTER 8
Control Systems

Before You Begin

Think about these questions as you study the concepts in this chapter.

- What are some pros and cons of open and closed loop control systems?
- Why is feedback important in a system?
- What are some pros and cons of using digital and analog devices?
- Why would writing a flowchart before writing a program be beneficial?

Open or Closed Loop?

Objective

The purpose of this activity is to practice identifying open and closedloop systems.

Procedure

Identify the following examples in Table 8-1 as either closed loop or open loop by marking a "C" or an "O," respectively, and explain why.

	Example	C or O?	Why?
1.	Furnace		
2.	Water sprinkler		
3.	Toaster		
4.	Video game		
5.	Microwave		
6.	Grammar correction in a word processing program		
7.	Windshield wipers		
8.	Elevator		
9.	Outdoor lights on timers		
10.	Seatbelt alarm in car		

TABLE 8-1 *Closed vs. Open Loop.*

Exercise 8.2 Practice Drawing Flowcharts

Objective

The purpose of this activity is to practice drawing flowcharts.

Procedure

Select *one* of the topics below and create a flowchart describing each step of that process using the appropriate ISO and ANSI symbols. Refer back to the Activity 3.1.2a Flowchart Guide and the Flowcharting.ppt document if needed. Your instructor will specify if the flowchart may be written or created in a Microsoft Word document and attached to this sheet.

1. Back your car out of the garage and drive to the nearest burger restaurant.
2. Make a pizza.
3. Give a friend directions to the mall.
4. Change the oil in a car.
5. Describe the process with which ring-shaped cereal pieces are extruded and made.

Exercise 8.3 Determining the Feedback Within a System

Objective

The purposes of this exerciseare to determine whether a hair straightener is an example of a closed-loop or an open-loop system and to draw a flowchart of its operation.

Materials

- Kil-A-Watt meter
- Hair straightener

Procedure

1. Plug the Kil-A-Watt Meter into a wall outlet.
2. Plug the hair straightener into the Kil-A-Watt Meter plug.
3. Turn on the hair straightener. If the straightener has heat settings, turn it to the highest setting.
4. Record the power ratings at 1-minute intervals in Table 8-2.
5. Be sure to turn off and unplug the hair straightener and let it cool completely before storing.

Safety Warnings

Hair straighteners can get extremely hot, and caution should be used with them. Many straighteners operate at 450°F to 500°F. Do not come in contact with the hot plates on the straightener. Place the hair straightener on a cork board or a heat-resistant surface.

Time (min)	Power (W)
0	
1	
2	
3	
4	
5	
6	
7	
8	
9	
10	

TABLE 8-2 *Power Ratings.*

Analysis

1. Describe the change in power used by the hair straightener beyond the 10-minute interval. Did it constantly increase? Did it decrease?

2. Construct a line graph showing the power used as a function of time using either graph paper or Microsoft Excel. Be sure to include a title and axis labels.

Conclusion

1. Is a hair straightener an example of a closed-loop or an open-loop system? How do you know?

2. Draw a flowchart using the appropriate ISO and ANSI symbols for the operation of a hair straightener. Refer back to the Activity 3.1.2a Flowchart Guide and the Flowcharting. ppt document if needed. Your instructor will specify if the flowchart may be written or created in a Word document and attached to this sheet.

CHAPTER 9
Materials

Before You Begin

Think about these questions as you study the concepts in this chapter.

- What are common domestic and engineering applications of both ferrous and nonferrous metals?
- How do softwoods differ from hardwoods, and what are the common applications of both materials?
- What are different categories of ceramic materials, and what kinds of products are made from materials in these categories?
- How do thermoplastics differ from thermosetting plastics, and what kinds of applications are these plastics used in?
- What is the difference between a matrix and reinforcement with a composite material, and what common applications use composites?

BACKGROUND

Categories of Materials

Materials for design and manufacturing can be artificial or naturally occurring. Unique materials can be created by combining two or more known materials. When creating a new material, the goal is to create a superior-performing item to meet the needs of the customer.

Materials fall into two broad categories: metallic and nonmetallic. These two categories are further broken into a variety of subcategories. These subcategories make up the structure of products today.

For example, let's look at the evolution of the humble ice cube tray (Figure 9-1). The first documented ice cube trays were made of stainless steel in the 1930s. Later came a rubber version. In the 1950s, aluminum trays were developed with a lever system to help loosen the cubes after freezing. A flexible plastic tray was developed in the 1970s. Today, there are silicone ice cube trays and built-in ice cube makers and dispensers.

Materials were important factors that influenced design improvements to the ice cube tray over time. Some people may remember getting their fingers stuck to aluminum ice cube trays and the ice sticking to the lever system, resulting in crushed ice, pain, and frustration.

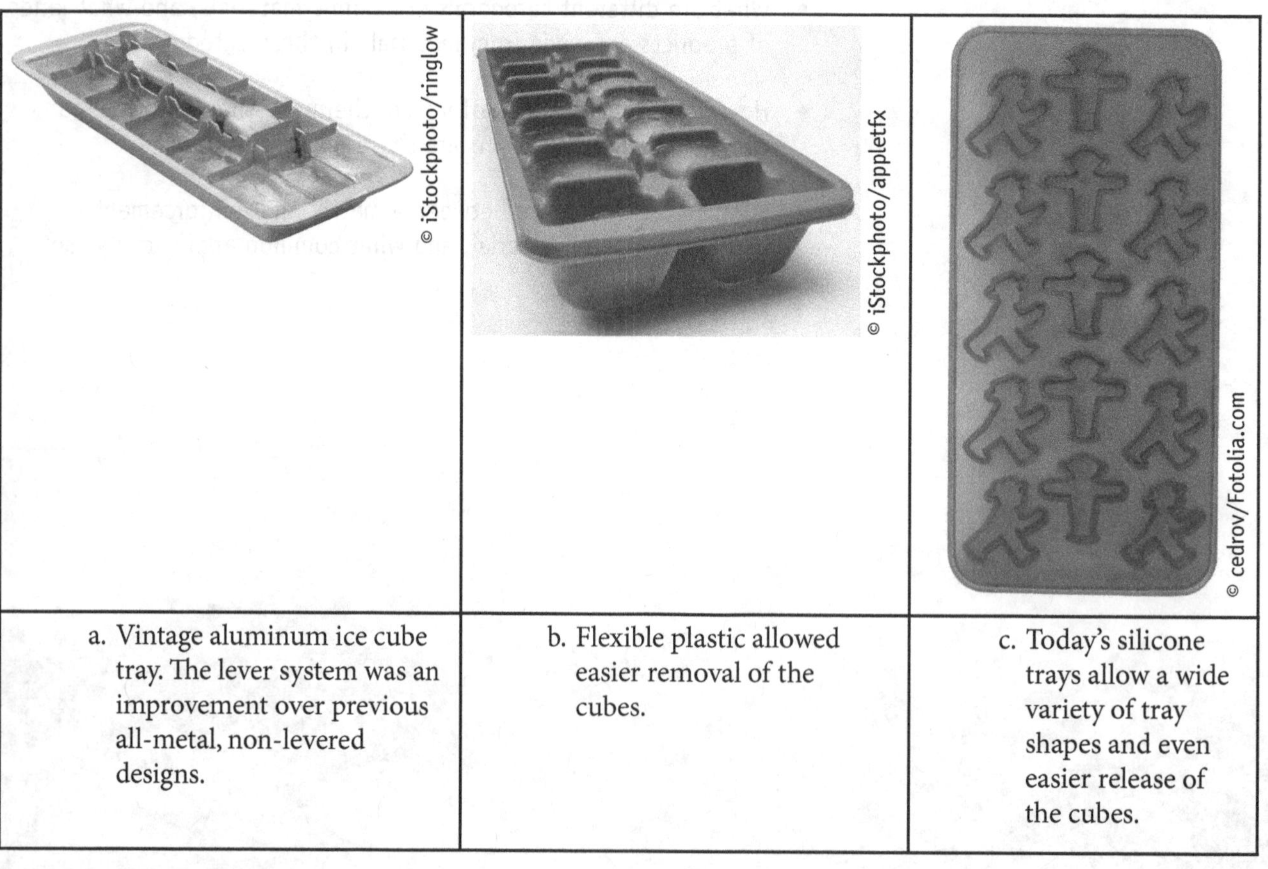

a. Vintage aluminum ice cube tray. The lever system was an improvement over previous all-metal, non-levered designs.

b. Flexible plastic allowed easier removal of the cubes.

c. Today's silicone trays allow a wide variety of tray shapes and even easier release of the cubes.

FIGURE 9-1 *Ice cube tray designs have evolved to take advantage of newer materials.*

Flexible plastic trays worked by popping whole cubes out of the tray, which was much easier than manipulating aluminum trays. However, the plastic trays tend to fatigue over time and crack as you flex them to remove the cubes. The use of silicone as a design material eliminated this problem without sacrificing ease of use.

The newest designs for ice cube trays are made from silicone. These trays are even more flexible and make it easier to get the cubes out of the tray. Because the silicone material stretches and flexes more than plastic or aluminum, designers can offer a wide variety of shapes, well beyond the basic cube.

The alphabet of materials in Figure 9-2 shows how materials are organized into subgroups based on their material properties.

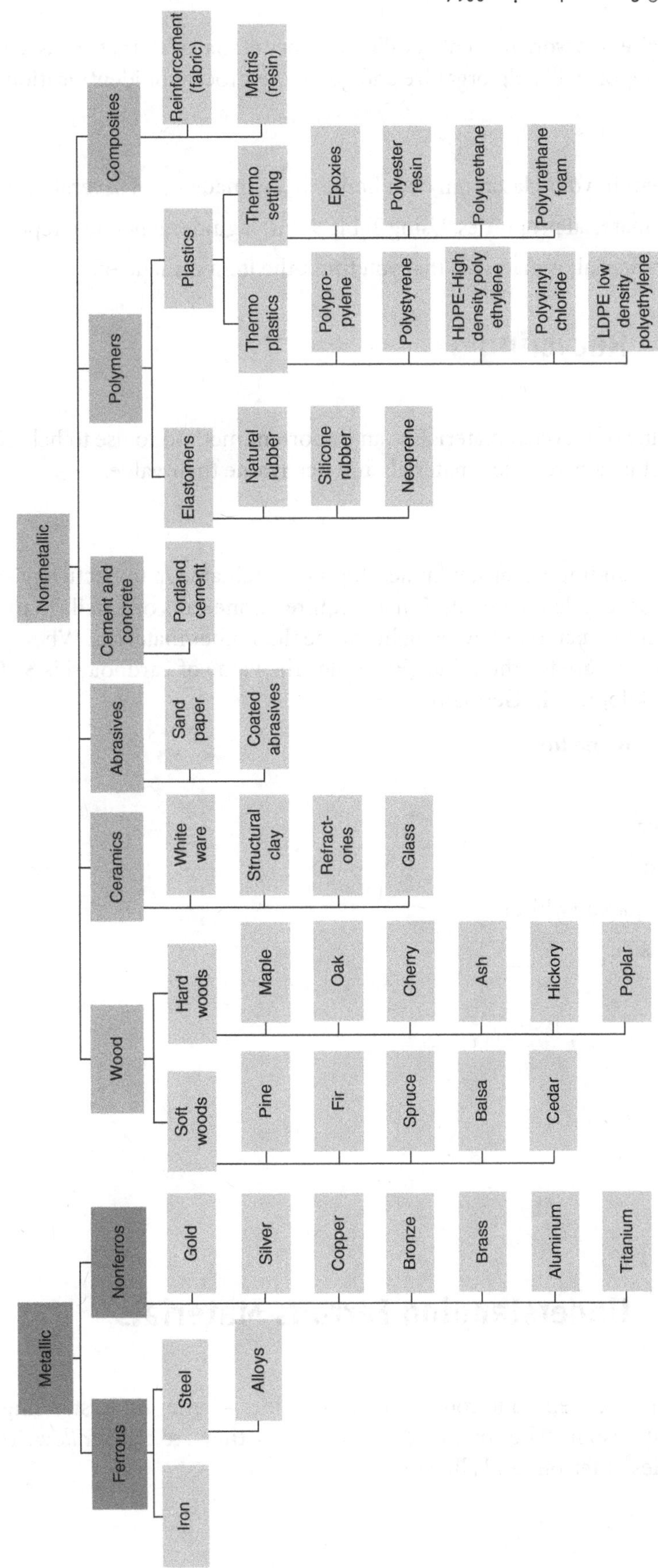

FIGURE 9-2 *The alphabet of materials.*

Exercise 9.1 Material Identification

Objective

The goal of this exercise is to work to identify different types of materials that are used in or on objects. Applying the material inspection report will help organize and guide the process of identification.

Procedure

1. Choose an item in your classroom or at home that is made up of several parts.
2. Describe the material properties. Using Table 9-1 as a guide, generate a report on the item.
3. Identify the material or materials that you think the item is made of.

Exercise 9.2 Recycling

Objective

Determining the value of recycled materials is an important method to use to help recover costs in manufacturing. Research a list of locally recycled materials and determine their value.

Procedure

Recycling materials is an important step in developing sustainable manufacturing systems. Research the values of the following list of recycled materials. It may require phone calls or emails to the local recycling companies. Try to indicate why some materials have a higher value than other materials. When providing pricing, make sure to identify a unit of measure for the value. (Example: The value of cardboard is $.005 per pound in the United States and euros per kilogram in Germany.)

- **Mixed plastic by the ton**
- Paper
  - Newspaper
  - Cardboard
  - Computer paper (white)
- High-carbon steel
- Copper
- Aluminum
- Brass
- Iron
- Bronze
- Platinum
- Titanium

Exercise 9.3 Understanding Ferrous Materials

Objective

Discovering the properties behind ferrous materials provides a better understanding of the spectrum of materials. Metallic materials also may be combined; in such cases, they are called *alloys*. The goal of this exercise is to identify the properties of ferrous and alloy materials.

Date:	Student Name:	Class:

Overall Item Description:

____

____

____

____

Material of Item 1 Material name	Description of properties	Picture
Material of Item 2 Material name	Description of properties	Picture
Material of Item 3 Material name	Description of properties	Picture
Material of Item 4 Material name	Description of properties	Picture
Material of Item 5 Material name	Description of properties	Picture
Material of Item 6 Material name	Description of properties	Picture

TABLE 9-1 *Material Inspection Form.*

Procedure

1. **Ferrous Materials**

   a. Name several characteristics of ferrous materials.

      i. ______________________________

      ii. ______________________________

      iii. ______________________________

      iv. ______________________________

      v. ______________________________

   b. What do ferrous materials have in common?

   c. Name some typical uses of ferrous materials.

2. **Alloys:** Using Table 9-2 as a guide, research and list six different metallic alloys. Include a product or a process use for each alloy.

Alloy	Product Use	Process Use
1.		
2.		
3.		
4.		
5.		
6.		

TABLE 9-2 *Metallic Alloys.*

3. Describe the difference between iron and steel.

4. Describe the difference between low-, medium-, and high-carbon steel.

5. **Corrosion or Oxidation:** Corrosion or oxidation can affect the performance of ferrous materials and alloys. Choose three ferrous metals and three metal alloys and identify the color of the corrosion. Record your findings in Table 9-3. Include a photo wherever possible. (If you like, you can recreate this table in a word processing program or spreadsheet and add a digital image.)

Material Name	Type(Ferrous/Alloy)	Corrosion/Oxidation	Color	Photo
1.				
2.				
3.				
4.				
5.				
6.				

TABLE 9-3 *Corrosion.*

6. **Melting Point:** Research the melting points of the following materials and record them in Table 9-4.

Metal	Melting Point(°F or °C)
Low-carbon steel	
Medium-carbon steel	
High-carbon steel	
316 stainless steel	
304 stainless steel	
Cast iron	
A2 steel alloy	
303 stainless steel	
A36 steel alloy	
4140 steel alloy	
1018 steel alloy	
Iron	

TABLE 9-4 *Melting Points of Ferrous Materials.*

Exercise 9.4 Understanding Nonferrous Materials

Objective

Discovering the properties behind nonferrous materials provides a better understanding of the spectrum of materials. This exercise develops your knowledge of nonferrous materials.

Procedure

1. Describe the characteristics of nonferrous materials. What do nonferrous materials have in common? What are typical uses of nonferrous materials?
2. Nonferrous materials have intrinsic value. For each of the nonferrous metals listed below, research and list the current selling prices at the New York Mercantile Exchange, the London Metals Exchange, or the Chicago Mercantile Exchange.
   a. Gold
   b. Silver
   c. Copper
   d. Palladium
   e. Platinum
3. Which of these nonferrous metals (gold, silver, copper, palladium, and platinum) has increased in price the most on a percentage basis over:
   The last year ______________________
   The last two years ______________________
   The last five years ______________________

4. Research the melting points of the following materials and record them in Table 9-5.

Metal	Melting Point(°F or °C)
Gold	
Silver	
Copper	
Lead	
Tin	
Titanium	
Zinc	
Cobalt	
Brass	
Bronze	
Molybdenum	
Aluminum 6061	
Tungsten	

TABLE 9-5 *Melting Points of Nonferrous Materials.*

Exercise 9.5 Understanding Wood Materials

Objective

Discovering the properties behind wood materials provides a better understanding of the spectrum of materials. This exercise develops your knowledge of wood materials.

Procedure

1. List five consumer products that are wood-based. What type of wood is each product made of? For each product, why do you think that type of wood was selected for that product?

   a.

   b.

   c.

   d.

   e.

2. Wood types are often subcategorized according to the variations found in their grain patterns. For example, maple variations include burl, spalted, wavy, tiger, flame, and quilted. Identify a wood product in a variation that you find visually appealing and explain why in a paragraph. (For example, search the Internet for the term "flame maple guitar" and see what images you can find.)

3. Like all materials, wood prices change based on supply and demand. Select six different wood species and research their prices. The price for each species should be based on a board-foot cost of an 8/4 (eight-quarter) piece of material, which is 2" thick. Figure 9-3 shows an example of calculations of board-feet.

4. The price of wood is calculated in board-feet. A board-foot is a 12" wide by 12" long by 1" thick (4/4, or four-quarter) piece of wood. So a piece of wood that is 8" wide by 7' long by 1" thick measures 4.66 board-feet. A board that is 5" wide by 14" long by 2" thick measures 0.97 board-feet.

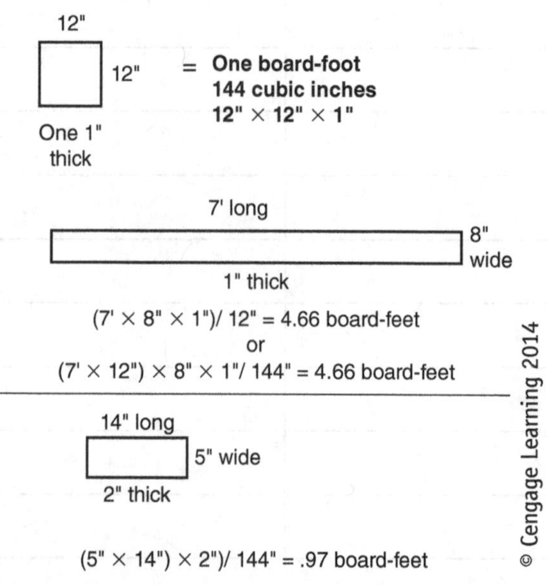

FIGURE 9-3 *The price of wood is calculated in board-feet.*

5. Calculate board feet for the wood measurements in Table 9-6.

Thickness	Length	Width	Total Board-feet
4/4	8'	7"	
4/4	30"	8.75"	
4/4	128"	14.5"	
8/4	4.5'	1.375'	
8/4	18"	8"	
8/4	12'	4.875"	
5/4	10'	16"	
5/4	48.75"	11.5"	
6/4	8.75'	1.3'	
6/4	28"	8"	
3/4	78.25"	1.5'	
3/4	15'	8.75"	

© Cengage Learning 2014

TABLE 9-6 *Board-Feet Calculation.*

Exercise 9.6 Using Abrasives

Objective

This exercise develops your understanding of how abrasives are used to remove material. They are also used to smooth the surface on a material to prepare it for finishing. There are many finishing items.

Procedure

1. In a team setting (teams of 2–4):
   a. Research different types of abrasive products and what types of materials they are made of.
      i. Gather different grades and types of abrasives (either real or just photographs).
      ii. Samples of abrasives can come in a variety of different forms (sandpaper, sanding blocks, diamond cutters, abrasive cutting discs, etc.).
   b. Develop a display board or set of sample boards showing the difference in the level of abrasive finishes. This can be using photographs of the objects or the real objects if acquired.
2. Develop a chart to explain the principles behind abrasive grit. Research the following items to gain a better understanding of how abrasive products are made:
      i. Explain the difference between the grit numbers that are used to categorize abrasives.
      ii. Explain the ranges of grit (coarse, medium, fine, or extra fine).
      iii. Research the most common materials used for the grit on abrasive products.
      iv. Find a video about how abrasive products are made and provide the link, if applicable.

Exercise 9.7 Understanding Ceramics

Objective

This exercise will help identify ceramic materials, which have unique properties that are used in both consumer products and manufacturing.

Procedure

1. Research and make a list of ceramic materials.
2. You might have noticed that some china plates break very easily, while others are very durable. Explain why.
3. Is the fiberglass insulation used in homes actually made from glass?
4. What makes fiberglass insulation so itchy?
5. Refractory ceramics were developed about 300 years ago. What properties make this process so relevant that it is still used today?
6. Explain the difference between plate glass and tempered glass.
7. A variety of coatings can be applied to glass. Research three of these coatings and write a paragraph about the uses of the coating.
   a.
   b.
   c.

Exercise 9.8 Understanding Cement and Concrete

Objective

This exercise helps you understand that cement and concrete are in the same family of products.

Procedure

1. What products are used to make concrete? What are the proper proportions for mixing concrete?
2. Concrete has a PSI rating associated with it. What does PSI stand for, and why is the rating significant?
3. Develop a matrix or table listing the major PSI ratings for concrete products and their uses.
4. What is the chemical makeup of Portland cement, and where is each of the components mined?

Exercise 9.9 Curing Concrete

Objective

Identify and describe the factors influencing the curing time of concrete. Students will add various additives to a concrete mixture and observe changes in the curing time.

Materials

- Gloves (made of nitrile)
- Plastic cups
- Mixing vessel for the class
- Water
- Portland cement
- Sand
- Handheld hair dryer
- Aggregates
- A variety of additives (whatever you think could influence the time to cure), such as the following:
    - Soda
    - Flour
    - Sugar
    - Cotton balls
    - Cookies
    - Additional water

Note:

Aggregates *are items that the concrete can bond to.*
Additives *are typically liquids or materials that influence the curing process.*

TIP SHEET

- This exercise will require an area that can get messy. If possible, cover the work area with plastic and tarps.
- *Note on material disposal*: Let the concrete cure before disposing of it in a trash receptacle.
- *Warning:* Do NOT put concrete down any sink. After handling concrete mixtures, wash your hands in a bucket of water that can be disposed of outside, and then wash your hands with soap in a sink.

Procedure

Put on protective gloves and eyewear.

1. Mix a small batch (a 1- to 40-lb bag of concrete) in a tub or wheelbarrow using a formula that you researched. Create a control sample in a 1-quart container by filling a plastic cup. Set your control sample aside to dry.
2. Record the curing time in 5-minute intervals using Table 9-7. Note the time it takes the sample to skin over and to hold a pencil vertically.
3. Using the same batch of concrete after you take the control sample, use quart containers of the concrete to introduce your selected additives to change the chemical composition in the plastic cups. Again, record the curing time in 5-minute intervals using Table 9-8, noting the skin-over time and the time to hold a pencil vertically.

TIP SHEET

You can use a hair dryer to speed up the curing process. You also can measure the effect of heat and air movement on the curing time.

4. Test the samples to see if any are stronger due to the additives.
5. Test each sample for hardness by using the end of a pencil eraser and depressing the sample, and then record the amount of resistance and the amount the material reflexes back to the original shape. Develop conclusions based on your data about the curing times and the strengths of the samples.

Time Interval	Temperature(°F or °C)	Hardness
5 minutes		
10 minutes		
15 minutes		
20 minutes		
30 minutes		

TABLE 9-7 *Curing Times.*

Time Interval	Compound A		Compound B	
5 minutes				
10 minutes				
15 minutes				
20 minutes				
30 minutes				

TABLE 9-8 *Curing Times—Compound A and Compound B.*

Exercise 9.10 Understanding Polymers

Objective

Polymer recycling is a huge part of the recycling effort in the United States. Each polymer product that is recyclable has a stamp that identifies it. This exercise will help you identify different polymer products.

Procedure

1. Collect three plastic materials with different recycling stamps on them. (Rinse any materials out of the container before bringing it to school.)
2. Research the different stamps and the product materials that go with them.
3. Make a slide presentation on each of the recycling stamps and the type of material it is made of. Each slide should have:

   a. A picture of the product

   b. An image of the stamp

   c. The type of material it is (based on the symbol research)

Exercise 9.11 Working with Polymers

Objective

The exercise helps you understand the effects of the creation of a polymer-based product. Polymers react in a variety of ways. For example, epoxy putty uses a chemical reaction to create a bonding agent that can hold materials together.

Materials

- Nitrile gloves
- Safety glasses
- Epoxy putty (comes in two parts that are mixed together to create a bonding agent)
- Paper
- Thermometer

Procedure

1. Form teams of two to three students.
2. Put on protective gloves.
3. Remove the plastic covering the epoxy putty.
4. Create equal portions of the two different putties. Do not use all the putty, and do not let the two putties touch until the next step.
5. Combine the two putties by kneading the putty until it is a single color. Note the time that it takes to combine the products. During the process of mixing, an exothermic reaction will take place.
6. Measure the temperate of the putty at 1-minute intervals.
7. Place the putty on a piece of paper and note the time.
8. Feel the putty as it sits on the paper. Does the putty release any heat?

9. Observe the putty's hardness after 1 minute, 5 minutes, 8 minutes, and 10 minutes. You can test for hardness by pressing on the putty with a gloved hand. Develop a table to record how much the putty hardens based on the following criteria:
   a. Record how easy the putty is to press with a gloved finger
   b. Do this at set intervals of time (every 15 or 30 seconds).
10. Partnering with another team, repeat the process using *unequal* portions of putty. Team will use more of putty 1, while team B will use more of putty 2, as follows:
   a. Team A can use double the amount of Part 1 of the epoxy and a single part of Part 2 of the epoxy.
   b. Team B can use double the amount of Part 2 of the epoxy and a single part of Part 1 of the epoxy.
11. Again, knead the putty until the color is consistent. Place the putty on the paper and look for changes in temperature and hardness.
12. Develop a table to record how much the putty hardens after 1 minute, 5 minutes, 8 minutes, and 10 minutes. Test the hardness by pressing on the putty with a gloved hand.
13. Answer the following questions for your team's putty compound:
   a. Did you notice any difference in the heat compared to your first, equal-portion mixture?
   b. Which mixture felt hotter?
   c. Which mixture felt colder?
   d. How long did it take for the each putty mixture to harden?
   e. Did the equal or the unequal portions harden fastest?

TIP SHEET

- Epoxy putty will stick to all surfaces, so make sure you place it on paper when working with it at all times.
- Wash hands immediately after working with the epoxy. If you accidentally get any epoxy on your skin, you can remove it with rubbing alcohol or an alcohol swab.

Exercise 9.12 Understanding Composites

Objective

Upon completion of this exercise, you will be able to understand and identify composite materials made of mixtures of two or more polymers.

Procedure

1. Describe the benefits and the drawbacks of using composite materials rather than alloy materials.
   a. Benefits:
   b. Drawbacks:
2. Research and provide three examples of a polymer matrix composite and a metal matrix composite. Enter the results in Table 9-9.

Polymer Matrix Composites	Metal Matrix Composites
a)	a)
b)	b)
c)	c)

© Cengage Learning 2014

TABLE 9-9 *Polymer vs. Metal Composites.*

3. Using 3-D design software, create the component shown in Figure 9-4. Calculate the surface area of the component.
4. Calculate the cost to create this part using sheets of 2-mm-thick solid carbon fiber.
5. Using the information in Figure 9-4, calculate the cost to make the part from aluminum.

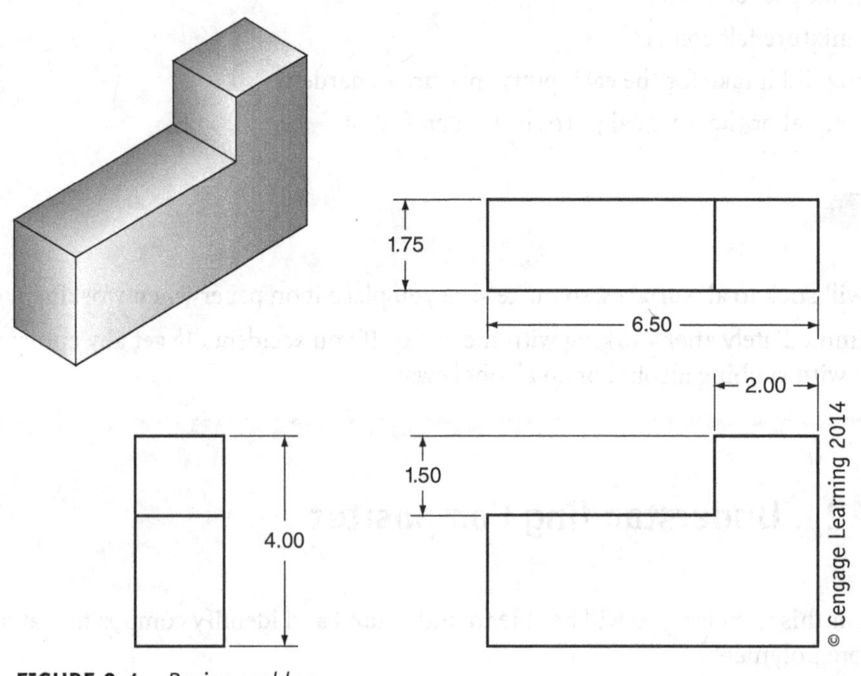

FIGURE 9-4 *Design problem.*

6. List at least three material properties for each of the following composite material categories. Enter the results in Table 9-10. (For example, high thermal shock resistance, chemical resistance, etc.)

Polymer matrix	a)
	b)
	c)
	d)
	e)
Metal matrix	a)
	b)
	c)
	d)
	e)
Ceramic matrix	a)
	b)
	c)
	d)
	e)
Carbon/carbon composites	a)
	b)
	c)
	d)
	e)

TABLE 9-10 *Material Properties.*

CHAPTER 10
Material Properties

Before You Begin

Think about these questions as you study the concepts in this chapter.

- What is the difference between stress and strain, and do they relate to a material's proportional limit?
- What are the proportional limit, yield stress, ultimate stress, and fracture stress, and where are they located on an engineering stress–strain diagram?
- Why do engineers apply factors of safety to their designs, and how do they use them to determine allowable stresses?

Explore Your World

The following website links apply to material properties as a whole:

- Found on the Matweb site are resources about specific material properties and values.

  Excellent resource on the properties of a variety of materials. **http://www.matweb.com**

- The Instron site is focused on the testing of materials and the standards that are used to apply the testing tools. By selecting a material, the different types of testing process are broken into the specific types of test that are used in gathering the data about the material.

  http://www.instron.us/wa/home/default_en.aspx

  Explanation of variables and processes used in material testing

Uniform Axial Stress

$$\text{uniform axial stress} = \frac{\text{normal force}}{\text{cross-sectional area}}$$

$$s = \frac{F}{A}$$

NOTE: At the end of the chapter, there is an equation sheet that lists the equation numbers.

Be careful—do not confuse pressure for stress. Pressure represents an external load, wherein a force acts across an object's surface, like wind on a building or on an object when it is submerged in water. Stress is a type of internal pressure that forms as a reaction to an external load.

Problem A sculpture is hanging from the ceiling from a tube of 2" in diameter; it weighs 200 pounds. What is the uniform axial stress applied on the tube?

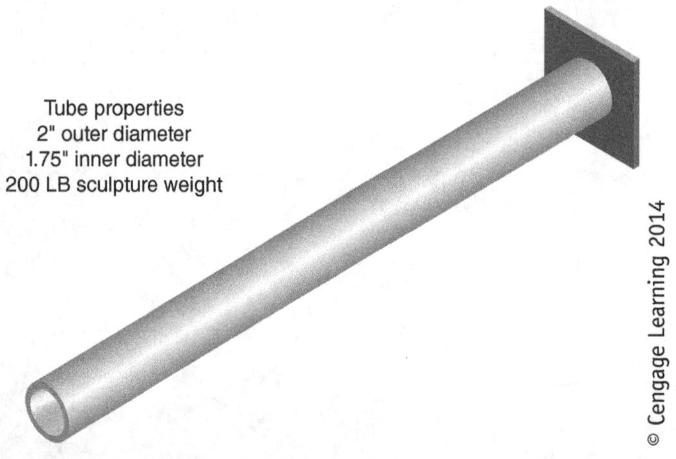

FIGURE 10-1 *Example of calculating uniform axial stress.*

Solution Step 1: Define the variables needed.

The cross-sectional area of the pipe must be calculated because it is not given in the problem.
Force involved: weight of the sculpture—200 lbs.
Step 2: Solve for the cross-sectional area of the pipe:

$$Area = \frac{\pi(d_{outside}{}^2 - d_{inside}{}^2)}{4}$$

d = the diameter of the pipe

Adding values:

$$Area = \frac{3.14[(2.0 \text{ in.})^2 - (1.75 \text{ in.})^2]}{4}$$

Condensing the equation:

$$Area = \frac{3.14[(2.0\text{ in.})^2 - (1.75\text{ in.})^2]}{4}$$

$$\frac{3.14[(4.0\text{ in.}^2) - (3.063\text{ in.}^2)]}{4}$$

$$Area = \frac{3.14(.938\text{ in.}^2)}{4}$$

$$Area = \frac{3.14(.938\text{ in.}^2)}{4}$$

$$Area = \frac{2.95\text{ in.}^2}{4}$$

$$Area = .737\text{ in.}^2$$

Neglecting the weight of the pipe, how much tensile stress exists within the section of the mounting pipe? Force is 200 lb (weight of sculpture):

$$stress = \frac{force}{area}$$

$$stress = \frac{200\text{ lb}}{.737\text{ in.}^2}$$

$$stress = 271.3\frac{\text{lb}}{\text{in.}^2}$$

Exercise 10.1 Calculating Axial Uniform Stress

Objective

Calculate the axial uniform stress for various objects.

Procedure

Calculate the axial uniform stress for the following objects in either tension or compression based on the indicated stress arrow.

TIP SHEET

Don't forget to calculate the cross-sectional area before calculating the stress.

Problem 10.1 Sizes are in inches.

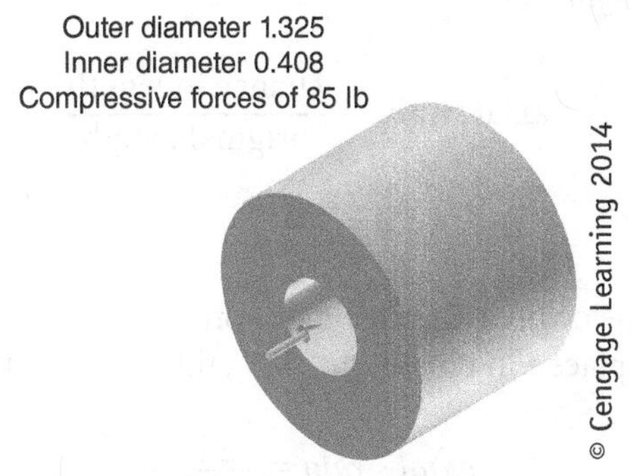

FIGURE 10-2 *Calculate uniform axial stress.*

Problem 10.2 Sizes are in inches.

FIGURE 10-3 *Calculate uniform axial stress.*

Problem 10.3 Sizes are in inches. Use the cylindrical area of the pin for calculation purposes.

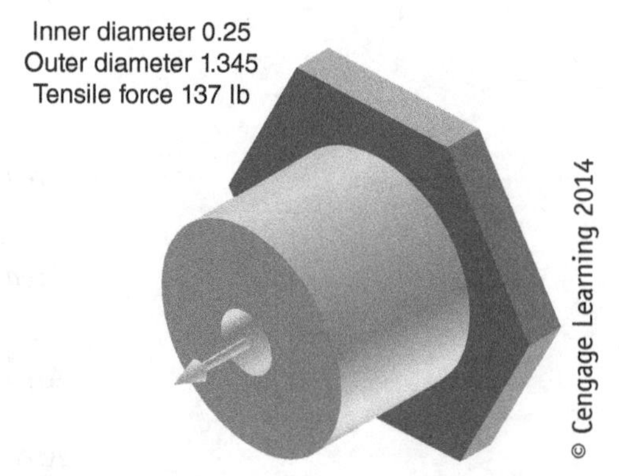

FIGURE 10-4 *Calculate uniform axial stress.*

Deformation and Strain

Strain comes in many forms, depending on the types of forces that are acting on an object.

$$\text{Change in Length} = \text{Final Length} - \text{Original Length}$$

$$\delta = L_f - L_o$$

Using licorice as an example, we will find the percentage of increase in length through axial strain. The ending measurement point of the strain should be where the object begins to break down in shape or split.

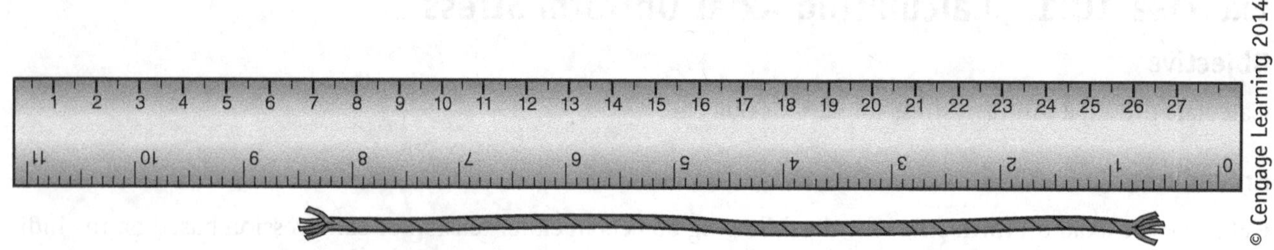

FIGURE 10-5 *Licorice at the standard length.*

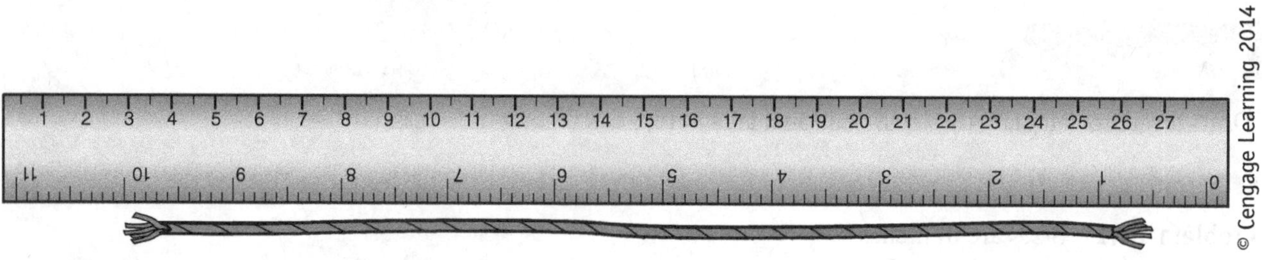

FIGURE 10-6 *Licorice at stretched length.*

$$\text{axial strain} = \frac{\text{change in length}}{\text{original length}}$$

$$e = \frac{\delta}{L_o}$$

When you pull on a piece of rope or licorice, the length of the item increases. So if a piece of licorice starts out 8.0" long and it stretches as it is placed in tension to 10.5", then the resulting change in length is 2".

$$\textit{axial strain} = \frac{2 \text{ in.}}{8.0 \text{ in.}}$$

$$\textit{axial strain} = .25$$

or a 25% increase in length of the licorice.

Exercise 10.2 Determining Axial Strain

Objective

Determine the axial strain of a piece of string or rope, licorice, or a rubber band.

Procedure

Determine the length of the object at rest between the locations where you will be holding the object (if possible, mark the locations with a pen). Then apply force to the object to stretch it, and have another person measure the distance of the object while it is in tension. Calculate the axial strain.

Conclusion

If you tried this exercise with a rubber band and it broke, then the following diagrams will have some meaning. They are stress–strain diagrams that illustrate what you did with the rubber band. By applying stress, the resulting strain (deformation or shape change) stretched the rubber band to the point that it broke.

Engineering Stress–Strain Diagram A graph that shows the relationship between stress (vertical axis) and strain (horizontal axis) for a specific material, based on a constant cross-sectional area. Think of the stress–strain diagram as a way to chart how a material reacts to force that is applied (stress) and the resulting deformation of the sample (strain).

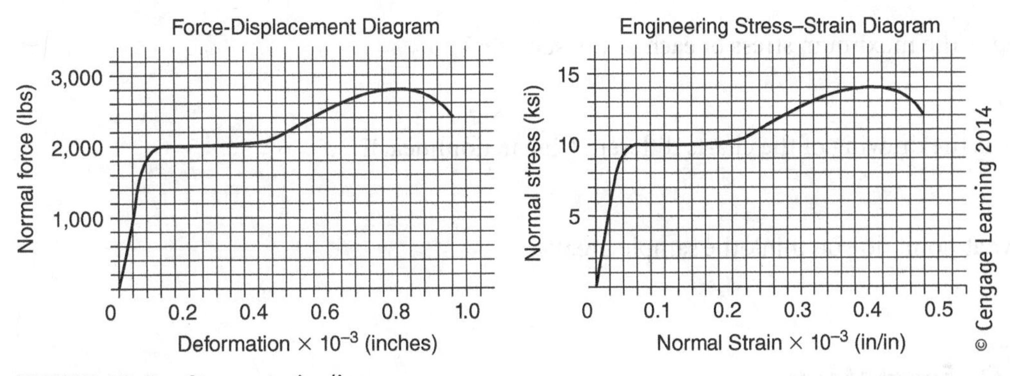

FIGURE 10-7 *Stress–strain diagrams.*

Force-Displacement Diagram A similar diagram, the force-displacement diagram is based on the force applied (stress) and the displacement (deformational change or movement). A good example is what happens in an earthquake. The chart in Figure 10-8 shows what happens to the Earth in a stick-and-slip type of earthquake, where the Earth's tectonic plates rub against each other and then slip, causing an earthquake. This is a typical type of earthquake that happens in California along the San Andreas Fault line.

The earthquake chart shows the release of the stress that has built up, which is similar to the force-displacement chart in Figure 10-8.

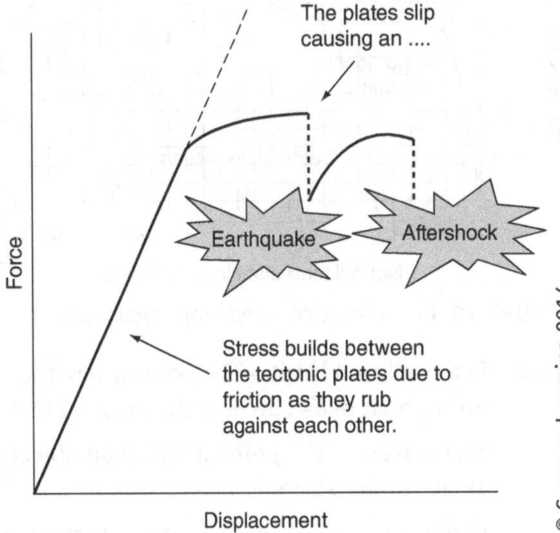

FIGURE 10-8 *Force-displacement diagram of an earthquake.*

Exercise 10.3 Stress–Strain Charts for Both a Metallic and a Nonmetallic Specimen

Objective

Compare the stress–strain for metallic and non-metallic specimens.

Procedure

Compare the charts and answer the following questions.

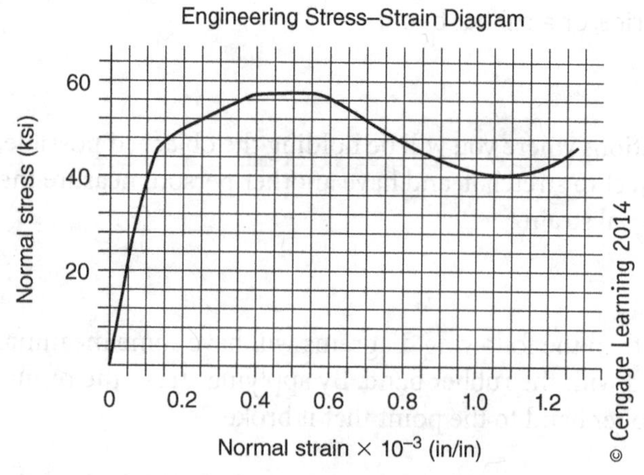

FIGURE 10-9 *Metallic stress–strain diagram.*

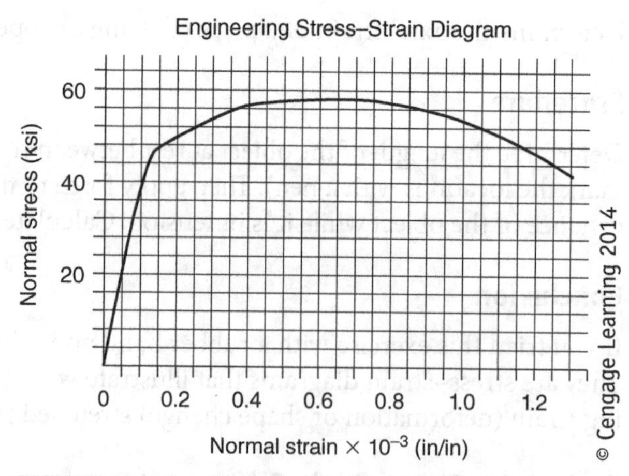

FIGURE 10-10 *Nonmetallic stress–strain diagram.*

1. What is similar about the charts?

2. What is the maximum stress of each of the samples?

3. How is the drawing of the charts different (linear, nonlinear)?

4. At what point do you think the sample breaks?

Elastic Deformation

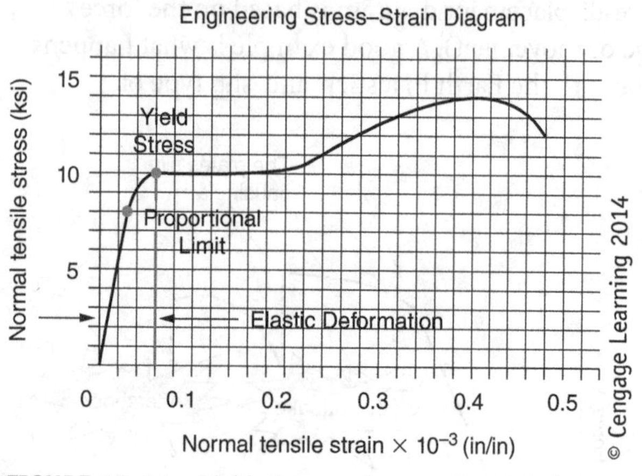

FIGURE 10-11 *Yield stress and proportional limit.*

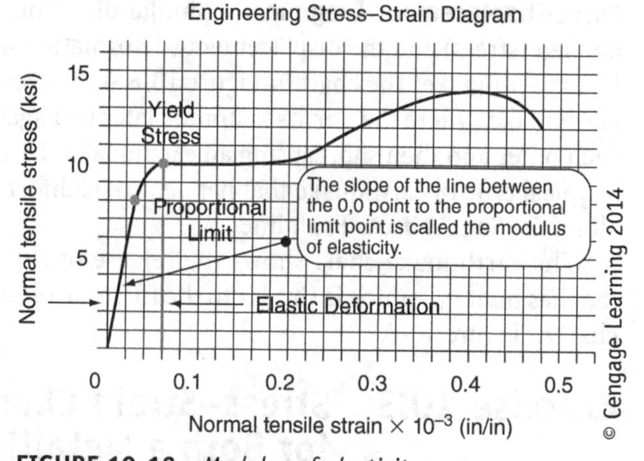

FIGURE 10-12 *Modulus of elasticity.*

- **Proportional limit**—The point where the elastic behavior of the sample no longer can be defined through Hooke's Law. It is the location that the stress is no longer proportional to the strain.
- **Yield stress**—The point at which an object will have a permanent change in shape once the sample is no longer under stress.
- **Elastic deformation**—The area under a curve that allows the object sample to return to its original shape once the stress is removed. The elastic deformation ends at the yield stress location on the chart.

- **Modulus of elasticity**—The slope of the line that falls from the start of the chart to the proportional limit is measured as the modulus of elasticity.

$$\text{modulus of elasticity} = \frac{\text{change in stress}}{\text{change in strain}}$$

$$E = \frac{\Delta s}{\Delta e}$$

Through his experiments, Robert Hooke discovered the law of elasticity, which he published in 1678. Today, we refer to Hooke's discovery as Hooke's Law.

$$\text{axial stress} = \text{modulus of elasticity} \times \text{axial strain}$$

$$s = Ee$$

When you think about the elasticity of some materials (especially metals), the amount of movement of the metal for a given force when it is applied is very small (i.e., it is difficult to "stretch" metal!). The modulus of elasticity would be extremely large due to the amount of stress that has to be applied to make the material strain.

The *Machinists Handbook*, discussed in the next exercise, has a complete listing of different materials' modulus of elasticity.

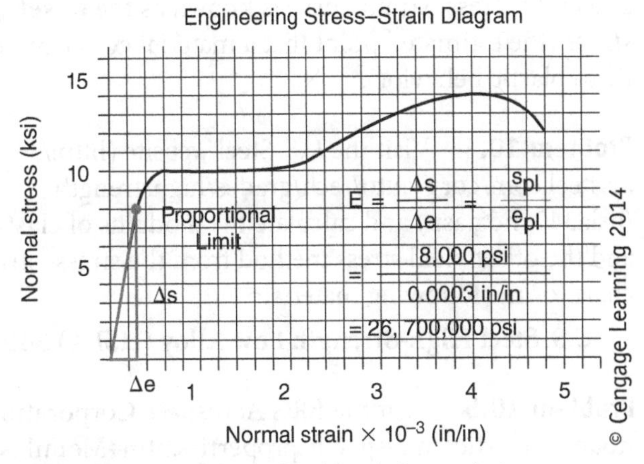

FIGURE 10-13 *Example of Hooke's Law being applied.*

Exercise 10.4 The *Machinists Handbook*

Looking at a chart on modulus of elasticity in the *Machinists Handbook*, you will find that each material has a different modulus of elasticity (E).

Objective

Identify the location where material property information can be gathered from trusted resources.

Procedure

Select two materials with different modulus of elasticity values and explain what the difference in the E values mean.

1. What does a higher or lower E value mean?
2. Is there any significance to the actual number difference between the two materials?

Exercise 10.5 Yield Stress

Yield stress is the point at which the resistance to maintain an object's shape gives way. Any attempt to return it to the original shape cannot succeed.

Objective

Identify the point at which the yield stress is found for the licorice piece.

Procedure

Using the licorice piece as a guide as you continue to pull on the licorice, you see it deform (change color) and elongate. Once the piece of licorice no longer returns to its original shape (about 2.0" longer than its original size), it indicates that you have exceeded the yield stress.

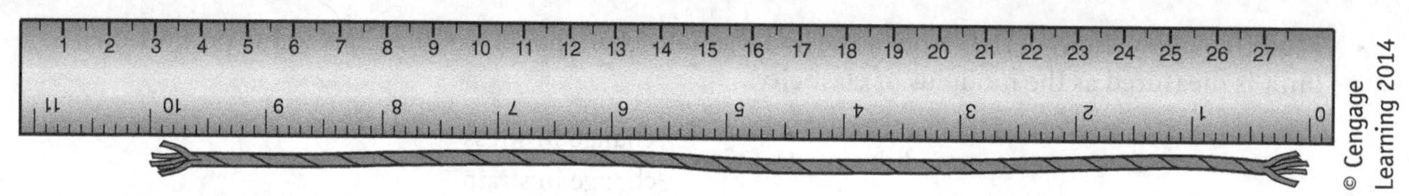

FIGURE 10-14 *Licorice stretched to yield stress.*

The standard starting point is 0.2% (0.002) of the normal strain on the x-axis. Drawing a line parallel to the straight-line portion of the diagram will develop an intersecting point to the diagram as the material becomes elastic.

The point where the offset parallel line crosses the engineering stress–strain curve is known as the **offset yield stress** (the estimated point that a material goes from elastic to plastic behavior).

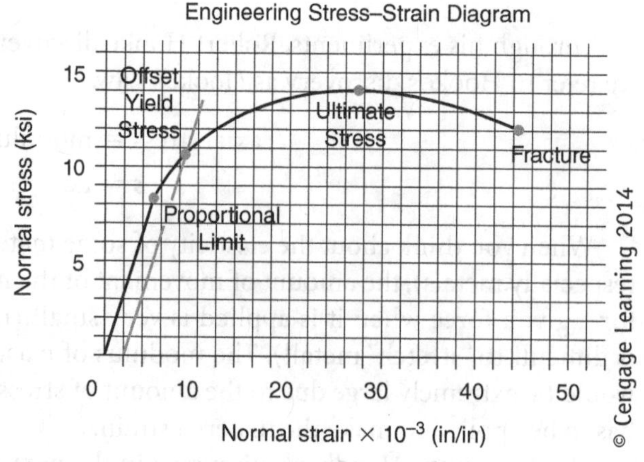

FIGURE 10-15 *Offset yield stress.*

Problem 10.4 Visit the U.S. Steel website (http://www.ussteel.com/corp/auto/tech/grades/highstrength/hsla340_hdg.asp) and calculate the modulus of elasticity and the offset yield stress method from the stress–strain chart for the following material:

U.S. Steel High-Strength Low Alloy (HSLA) 340 HDGI

Problem 10.5 Visit the ERG Aerospace Corporation's website (http://www.ergaerospace.com/foamproperties/matspecificproperties.htm#Modulus) and calculate the modulus of elasticity for a Duocel® foam product.

Determine strain from the modulus of elasticity chart on the website.
Stress value $\sigma = 6.50$ psi

Factors of Safety and Allowable Stress

$$\text{Allowable Stress} = \frac{\text{Yield Stress}}{\text{Factor of Safety}}$$

$$s_{\text{allowable}} = \frac{s_y}{n}$$

Why Are Safety Factors Important?

The safety factor is used in design to ensure that normal use of the product will not cause it to fail based on design criteria. However, failures could still occur due to manufacturing abnormalities, not using the product in a proper fashion, material defects, etc.

How much weight can a half-ton pickup hold? (1,000 lb) The frame and axles were tested to ensure that 1,000 lb can be hauled in the pickup safely. With a safety factor involved, it would hold up to 1,500 lb before complete failure of the frame and axles. Thus, you use a 1.5 safety factor to the original design criteria to calculate the design failure limit:

$$1{,}000 \text{ lbs} = \frac{\text{yield stress}}{\text{safety factor} = 1.5}$$

$$\textit{Yield stress} = 1{,}000 \text{ lb.} * 1.5$$

$$\textit{Yield stress} = 1{,}500 \text{ lb.}$$

Problem 10.6

1. Ladders generally have a safety factor of 4. Use the different types of ladder classes and their weight limits to calculate the yield stress of the ladder based on the weight limit of the ladder.

   Type III = 200 lb   Type IA = 300 lb

   Type II = 225 lb   Type IAA = 375 lb

   Type I = 250 lb

2. Why is a safety factor important for ladder usage?

Problem 10.7 A brass tube has an outer diameter of 1 in. and an inner diameter of 0.85 in. The tube is subject to a tensile load. The alloy material has a yield strength of 24,000 psi. The safety factor of this system is 2.50. Calculate the allowable stress and the tensile load.

Problem 10.8 The titanium connecting rod is under a tensile stress. The diameter of this solid rod is 0.375 in. With a safety factor of 3, determine the allowable stress and the tensile load.

Problem 10.9 The ladder rung is under a compressive load. This hollow aluminum 6061 rung has an exterior diameter of 1.25 in. and an interior diameter of 0.80 in. Using the standard ladder safety factor (in the reading), determine the allowable stress and the compressive load.

Percent Reduction in Area and Percent Elongation

Information about the change in diameter and the change in length of a fractured test specimen is used to calculate the *percent reduction in area* and the *percent elongation*. Together, these percentages indicate the degree to which a material exhibits brittle or ductile qualities. The **percent reduction in area** is a measure of the amount of necking, which is the decrease in the cross-sectional area where the fracture will take place.

$$\%\ \text{area reduction} = \frac{\text{original area} - \text{final area}}{\text{original area}}(100)$$

$$\%\ \text{area reduction} = \frac{A_0 - A_f}{A_0}(100)$$

The **percent elongation** is a measure of the amount of stretching that occurs within the gauge length of a tensile test sample. Mark the test sample, keeping the markings a standard distance apart. Then put the fractured sample back together as best you can and remeasure the distance between the two marks. The difference between the two measurements can be used to calculate the percentage of elongation:

$$\%\ \text{elongation} = \frac{\text{final gauge length} - \text{original gauge length}}{\text{original gauge length}}(100)$$

$$\%\ \text{elongation} = \frac{L_f - L_o}{L_o}(100)$$

Putting It All Together

Tensile stress, cross-sectional area, safety factor, % area reduction, and % elongation are the items that will be calculated in this example.

The sample of aluminum alloy has the following properties:

- Modulus of elasticity (E) = 15,000,000 psi
- 0.2% yield stress (sy) = 55 ksi
- Ultimate stress (su) = 63 ksi.

The test sample had an original length (L_o) of 1 in. and an original diameter (D_o) of 0.1875 in. After the sample

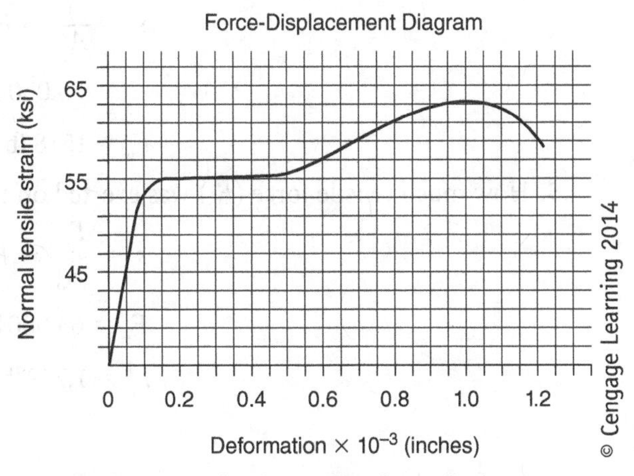

FIGURE 10-16 *Aluminum alloy force-displacement diagram.*

was tested in a tensile test machine, the final diameter (D_f) at the fracture point was 0.082 in. and the sample's final length (L_f) was 1.22 in.

1. What was the original cross-sectional area (A_o) of the tensile test sample within the 1-in. test length?

$$A_o = \frac{\pi\, d_o^{\,2}}{4}$$

$$A_o = \frac{3.1416 \times (0.1875\,\text{in.})^2}{4}$$

$$A_o = \frac{3.1416 \times 0.0352\,\text{in.}^2}{4}$$

$$A_o = \frac{0.1104\,.\text{in.}^2}{4}$$

$$A_o = 0.0276\,\text{in.}^2$$

2. What was the final cross-sectional area (A_f) of the tensile test sample?

$$A_f = \frac{\pi\, d_f^{\,2}}{4}$$

$$A_f = \frac{3.1416 \times (0.082\,\text{in.})^2}{4}$$

$$A_f = \frac{3.1416 \times 0.0067\,\text{in.}^2}{4}$$

$$A_f = \frac{0.0211\,\text{in.}^2}{4}$$

$$A_f = 0.0053\,\text{in.}^2$$

3. What was the percent reduction in the area of the tensile test sample?

$$\%\ \text{area reduction} = \frac{A_0 - A_f}{A_o}(100)$$

$$\%\ \text{area reduction} = \frac{0.028\,\text{in.}^2 - 0.005\,\text{in.}^2}{0.028\,\text{in.}^2}(100)$$

$$\%\ \text{area reduction} = \frac{0.023\,\text{in.}^2}{0.028\,\text{in.}^2}(100)$$

$$\%\ \text{area reduction} = 0.82 \times 100$$

$$\%\ \text{area reduction} = 82\%$$

4. How much tensile force (F_y) was exerted on the tensile test specimen at the 0.2% yield stress point?

$$s_y = \frac{F_y}{A_o} \Rightarrow F_y = s_y A_o$$

$$F_y = 55{,}000\ \text{lb/in.}^2 \times 0.0276\ \text{in.}^2$$

$$F_y = 1518\ \text{lb}$$

5. How much tensile force (F_u) was exerted on the tensile test specimen at the ultimate stress point?

$$s_u = \frac{F_u}{A_o} \Rightarrow F_u = s_u A_o$$

$$F_u = 63{,}000\,\text{lb/in.}^2 \times 0.0276\,\text{in.}^2$$

$$F_u = 1{,}739\,\text{lb}$$

6. What was the percent elongation of the tensile test sample at the point of fracture?

$$\% \text{ elongation} = \frac{L_f - L_o}{L_o}(100)$$

$$\% \text{ elongation} = \frac{1.22 \text{ in.} - 1.00 \text{ in.}}{1.00 \text{ in.}}(100)$$

$$\% \text{ elongation} = \frac{0.22 \text{ in.}}{1\,00 \text{ in.}}(100)$$

$$\% \text{ elongation} = 0.22 \times 100$$

$$\% \text{ elongation} = 22\%$$

Problem 10.10 Using the diagram below, complete the analysis of the following: tensile stress, cross-sectional area, safety factor, % area reduction, and % elongation.

The sample of nickel alloy has the following properties:

- Modulus of elasticity (E) = 26,000,000 psi
- 0.2% yield stress (sy) = 55 ksi
- Ultimate stress (su) = 63 ksi.

The test sample had an original length (L_o) of 1 in. and an original diameter (D_o) of 0.12 in. After the sample was tested in a tensile test machine, the final diameter (D_f) at the fracture point was 0.060 in.

Problem 10.11 Using the diagram below, complete the analysis of the following: tensile stress, cross-sectional area, safety factor, % area reduction, and % elongation.

The sample of nylon has the following properties:

- Modulus of elasticity (E) = 400,000 psi
- 0.2% yield stress (sy) = 18 ksi
- Ultimate stress (su) = 18 ksi.

The test sample had an original diameter (D_o) of 0.125 in. After the sample was tested in a tensile test machine, the final diameter (D_f) at the fracture point was 0.03 in.

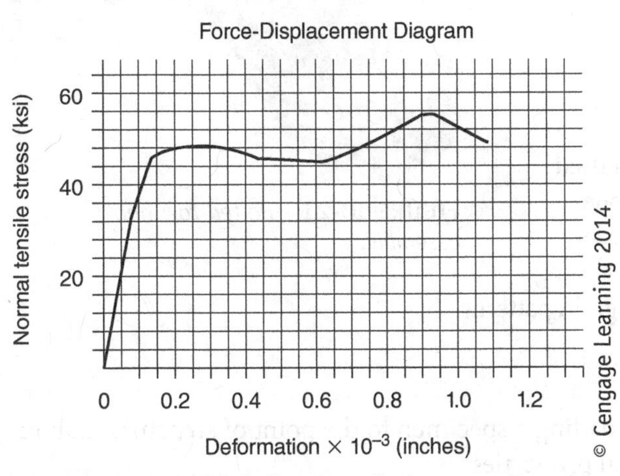

FIGURE 10-17 *Nickel alloy force-displacement diagram.*

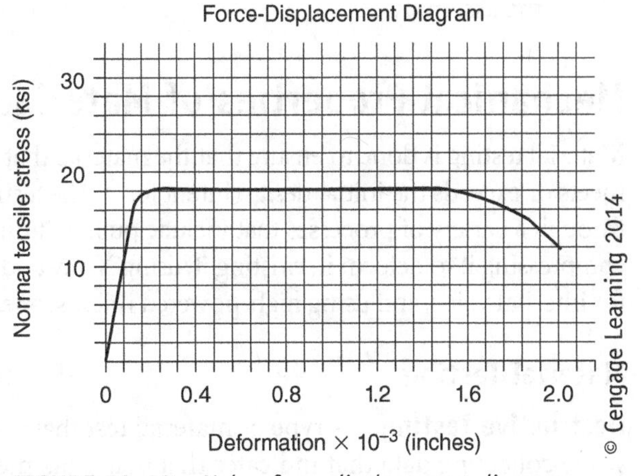

FIGURE 10-18 *Nylon force-displacement diagram.*

Elongation at break is 50%; original sample length (L_o) is 1.33 in.

Shear Stress

Mechanical fasteners (bolts, welds, rivets, pins, etc.) holding a structure together are subject to **shear force** (V), which generates a slicing action through the material as a result of **shear stress** (τ). Such connections must be designed to withstand shear stress without experiencing permanent deformation. In Figure 10-19, you can see this type of connection and the potential shear force.

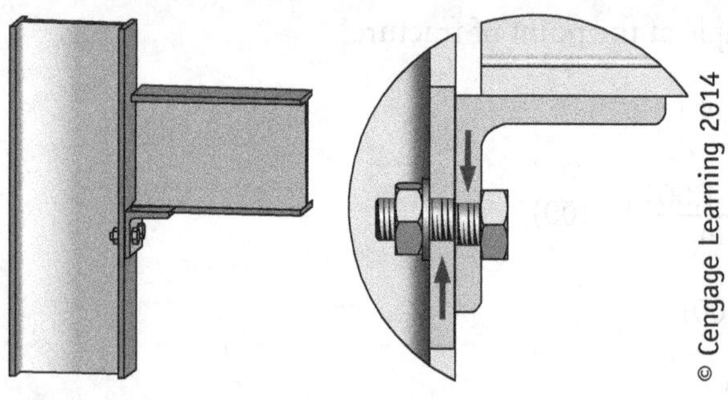

FIGURE 10-19 *Bolt shear due to stress.*

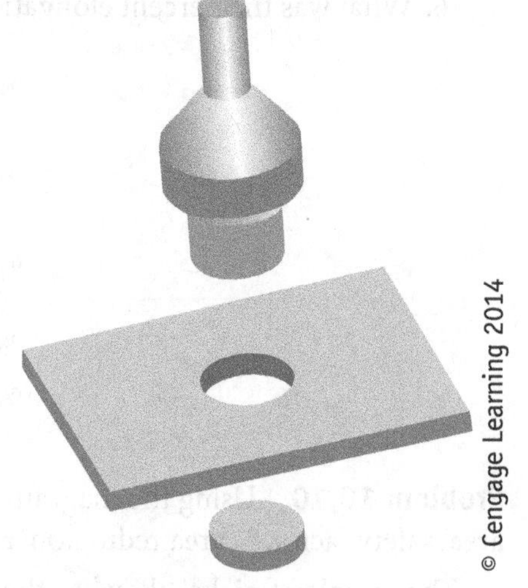

FIGURE 10-20 *Determine the shear force required to punch a hole.*

$$\text{Shear Stress} = \frac{\text{Shear Force}}{\text{Shear Area}}$$

$$\tau = \frac{V}{A_s}$$

Problem 10.12 A 1" diameter punch is used to cut a piece out of a 0.250"-thick piece of brass. The shear strength of the plate is 34,100 psi. The area is the shaded surface of the disk.

Find the shear force required to punch a hole like in the example shown in Figure 10-20.

Problem 10.13 A 0.375" diameter bolt is holding a tow hook tightly to the truck frame.

1. What is the shear stress of the bolt if the fastener is rated at 75,000 psi?
2. Determine the diameter of the smallest bolt possible to hold the assembly.

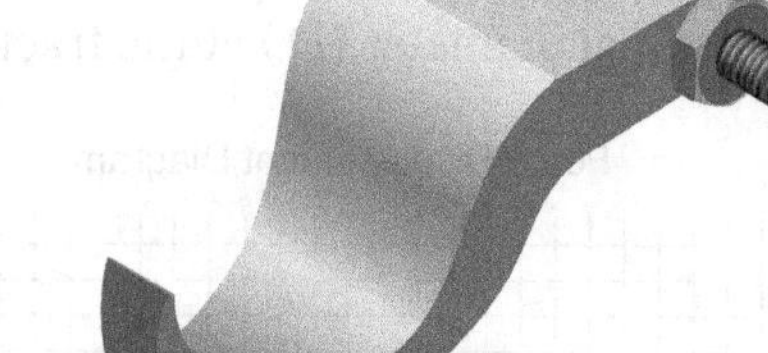

FIGURE 10-21 *Bolted tow hook assembly.*

Mechanical Properties of Materials

Material testing is done to ensure that the material that is being used meets or exceeds the initial design intent for it. The testing is done through a variety of processes that include pulling it until it breaks, compressing it to deform it, twisting it, using X-ray and MRI scanning for internal voids, and using high-powered microscopes to see grain patterns.

Material Testing

Destructive Testing A type of material test that involves loading a specimen to the point of structural failure while collecting data that indicates that material's mechanical properties.

Tensile Testing Woods, ceramics, and especially plastics and metals are placed through a common type of destructive stress test, called a *tensile test*, which measures a material's strength in tension.

Dynamic tensile tests are performed to determine how many times a material can be loaded before it will fail due to fatigue. Fatigue is the deterioration of a material as a result of repeated loading and unloading.

Use YouTube or Teacher Tube online video repositories to search the specific material testing processes and locate a video that shows how the test is done.

Compression Testing Some materials can withstand much larger compression stresses than tensile stresses. One example is concrete, which is used extensively in the foundations of buildings and bridges. Compression tests are also conducted on a universal testing machine. Compression samples are usually made in the form of cylinders that have a 2:1 length-to-diameter ratio.

Use YouTube or Teacher Tube online video repositories to search the specific material testing process and locate a video that shows how the test is done.

Shear Tests Fasteners and other cylindrical components that are subject to direct shear forces are often tested in a double-shear machine using specially designed fixtures on a universal testing machine. Use YouTube or Teacher Tube online video repositories to search the specific material testing process and locate a video that shows how the test is done.

Torsion Testing As mentioned earlier in this chapter, engineered objects like drive shafts, fasteners, and twist drills must withstand torsional loads that generate shear stresses. The materials from which such objects are made are subjected to a type of destructive test called a *torsion test*. Torsion tests provide the engineer with information about a material's modulus of elasticity in shear, yield stress in shear, ultimate shear stress, fracture stress, and other important information. Use YouTube or Teacher Tube online video repositories to search the specific material testing process and locate a video that shows how the test is done.

This formula is used to calculate the maximum shear stress on the outer surface of a solid cylindrical test sample at any point along the torque-angle curve:

$$\tau_{\text{max shear stress at the outer surface}} = \frac{16T}{\pi d^3}$$

Exercise 10.6 What Is Material Testing All About, and Why Is Testing Important?

Objective

Material testing is an important tool to ensure that the design object will perform as intended. This exercise identifies a variety of material testing processes and applies their calculations.

Procedure

1. Break into teams of two.
2. Select a material testing process.
3. Develop a classroom presentation on the process selected, using a variety of media applications (Microsoft® PowerPoint®, video, images, text, interviews, etc.).
   a. Research the testing process.
      i. What is the test used for?
      ii. What results will you gain from the test?
      iii. What are the steps in this process?
   b. Find or create a video or a series of images of the testing process.
   c. Report on an industrial use for the testing process.

Problem 10.14 Calculate the maximum shear stress for a shaft that has a diameter of 1.50 in. and a torque of 10,620 pound-feet.

Problem 10.15 Calculate the shear stress of a 50-mm-diameter shaft with a torque of 1,200 Nm.

Problem 10.16 Calculate the diameter of the shaft if the torque is 1,900 Nm and the shear stress is 0.08524 N/m^2.

Centroids and Area Moment of Inertia

A beam's cross-sectional shape is very important, as some shapes provide greater stability than others. The center of a shape is called a **centroid.** For a simple rectangular beam, the centroid is located at one-half of the base distance (b) and one-half of the height distance (h), as measured from a common origin point. A **centroidal axis** is an imaginary line that passes through the centroid and is appropriately aligned with one of the three principal Cartesian coordinate axes (*x*, *y*, or *z*) depending on the orientation of the structural member in three-dimensional space.

The idea that one shape may be stiffer in bending than another can be a bit confusing. The way that engineers can tell is through calculating a shape's *area moment of inertia.* **Area moment of inertia** (symbol I_{XX}), also referred to as the *second moment of an area*, is a geometrical property of a structural member's cross-sectional shape, which is measured in inches raised to the power of 4 (in.^{4}). This property reflects how the area of a shape is distributed in relation to its centroidal axis.

The area moment of inertia will indicate which shape has the highest capacity to resist rotating in the plane of its centroidal axis (such as the *x–x* axis). If the majority of the shape's area is distributed a large distance from the centroidal axis, then the shape will provide greater stiffness.

The area moment of inertia (I_{XX}) for a simple square or rectangular beam, given its base (*b*) and height (*h*) dimensions, is calculated as follows:

$$I_{XX} = \frac{bh^3}{12}$$

Problem 10.17 Calculate the area moment of inertia of a 6-in. by 4-in. piece of aluminum 6061.

Problem 10.18 Calculate the base size of a rectangular piece of steel when its height is 12.0 in. and the moment of inertia is 61.25 in.^{4}

Problem 10.19 Calculate the area moment of inertia of a 1.375-in. by 2.875-in. piece of bronze.

Problem 10.20 Calculate the area moment of inertia of a 9.276-in. by 3.367-in. piece of aluminum 6061.

Calculating Deflection

Several variables must be known in order to calculate the maximum deflection of a beam that experiences a point load at its midspan, as shown in Figure 10-22.

These variables include the modulus of elasticity (E) of the material that the structural sample is comprised of the span of the unsupported section of the flexural sample (*L*) in inches, the magnitude of the point load (*F*) in pounds, and the area moment of inertia (I_{XX}) of the sample's cross-sectional shape in in.^{4}.

This can be algebraically manipulated to calculate the modulus of elasticity for a material like wood, which naturally exhibits large variations due to the different circumstances under which trees grow and the different ways in which lumber is cut from a tree trunk.

$$\Delta_{max} = \frac{FL^3}{48EI_{XX}}$$

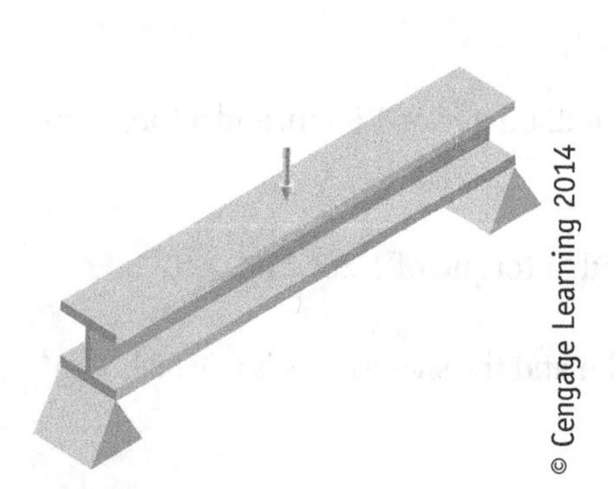

FIGURE 10-22 *Deflection calculation and point loading.*

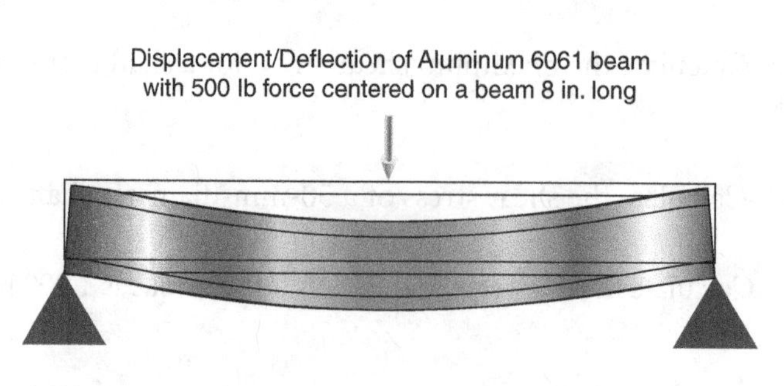

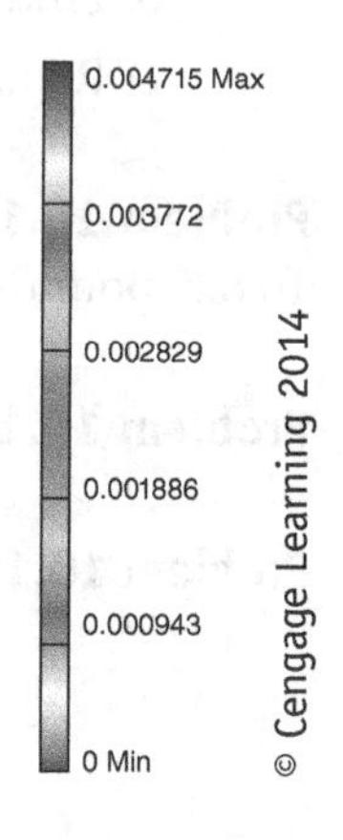

FIGURE 10-23 *Deflection of an aluminum I-beam.*

The following equation is used to calculate the flexural stress that is generated in the test specimen, which is measured in lb/in.^{2} :

$$s_{\text{flexure}} = \frac{3FL}{2bh^2}$$

Computer modeling can also be used to calculate the results, as seen in Figure 10-23.

Putting it All Together

Calculate the deflection of a 6" square, 6061 aluminum beam that has a clear span of 8 in. with a force of 500 lb centered on it.

1. Calculate the moment of inertia:

$$I_{XX} = \frac{bh^3}{12}$$

$$I_{XX} = \frac{(6\text{in.})6\text{ in.}^3}{12}$$

$$I_{XX} = \frac{1{,}296\text{ in.}^4}{12}$$

$$I_{XX} = 108\text{ in.}^4$$

2. Calculate the deflection:

F = Force or load

L = Length of span

E = Modulus of elasticity

I_{xx} = Moment of inertia

$$\Delta_{\max} = \frac{FL^3}{48EI_{XX}}$$

$$\Delta_{\max} = \frac{(500\text{ lb})(8\text{ in.})^3}{48(10{,}000\text{ lb/in.}^2)(108\text{ in.}^4)}$$

$$\Delta_{\max} = \frac{256{,}000\text{ lb-in.}^3}{51{,}840{,}000\text{ lb-in.}^2}$$

$$\Delta_{\max} = .00494\text{ in.}$$

Problem 10.21 Calculate the deflection of a 8" × 12" – 6061 aluminum beam that has a clear span of 48 in. with a force of 8,500 lb centered on it.

Problem 10.22 Calculate the deflection of a non-centered force on a beam that is 6" × 6" – 6061 aluminum beam that has a clear span of 60 in. with a force of 8,500 lb located 24" from the right side, as shown in Figure 10-24.

L = overall length

$$\Delta_{\max} = \frac{Fa^2\, b^2}{3EI_{XX}\, L}$$

Problem 10.23 Calculate the deflection of a 10" × 24" – 1025 low carbon steel beam, that has a clear span of 15 feet with a force of 5,500 lb centered on it.

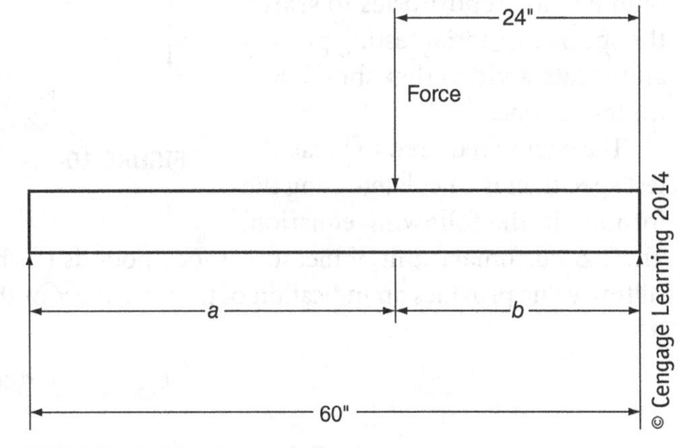

FIGURE 10-24 *Beam deflection.*

Problem 10.24 Calculate the deflection of a non-centered force on a 4" × 12" – polyethylene (HDPE) beam that has a clear span of 24 in. with a force of 300 lb 10" from the left.

Problem 10.25 Calculate the deflection of a 6" × 12" – cast iron (ASTM #50) beam , that has a clear span of 15 feet with a force of 13,000 lb centered on it.

Hardness Testing

To verify that a material has the proper hardness, it may be subjected to one of several types of hardness tests. Most of these tests involve indenting the material using a special shaped tool under a given static load, and measuring the resulting indent geometry. Three of the most common hardness tests are the Brinell test, the Rockwell test, and the Vickers test.

Use YouTube or Teacher Tube online video repositories to search the specific material testing process and locate a video that shows how the test is done.

Though the proper unit of measurement for a Brinell hardness value is kgf/mm^2, it is standard practice to identify the number as a unitless value, followed by initials that identify the type of hardness test that was used (BHN or HB).

$$HB = \frac{2F}{\pi D (D - \sqrt{D^2 - d^2})}$$

Toughness and Impact Strength Testing

The Charpy impact test is used to determine the amount of energy needed to fracture a specimen. The standard impact velocity for a Charpy impact test is 17.5 ft/s. Energy from the moving pendulum is transferred to the test specimen upon impact, wherein the sample fractures and the pendulum continues to swing past the point of impact. The final angle (symbol β, pronounced "beta") on the other side of the scale is recorded by a marker where the pendulum stops swinging. The kinetic energy (E) of the pendulum is a function of

- The weight of the pendulum (w) in pounds
- The length of the pendulum (r), as measured from the fulcrum to the center of the pendulum's mass in feet
- The pendulum's initial and final angle in degrees

Use YouTube or Teacher Tube online video repositories to search the specific material testing process and locate a video that shows how the test is done.

The energy required to break the test specimen is calculated using the formula in the following equation. The U.S. customary unit of measure is foot-pounds (ft-lb) and the metric unit is Newton-meters (N·m). The resulting value provides an indication of the toughness of the material through its energy absorption capacity.

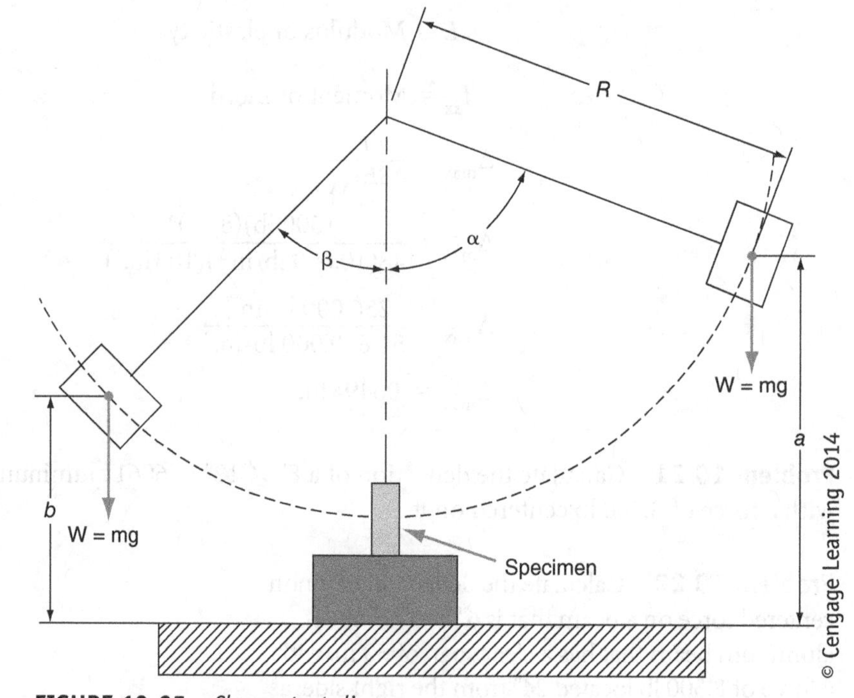

FIGURE 10-25 *Charpy test diagram.*

$$E_{Charpy} = wr(\cos\beta - \cos\alpha)$$

Problem 10.26 An impact testing machine has a hammer weighing 75 lb, with a drop height of a = 40 in. The length of the striking arm is r = 35 in. The sample's notch angle is 45°, and the pendulum's angle of rise of 85°.

Equation Sheet

Uniform Axial Stress

$$\text{uniform axial stress} = \frac{\text{normal force}}{\text{cross-sectional area}}$$

$$s = \frac{F}{A}$$

Change in Length

$$\text{change in length} = \text{final length} - \text{original length}$$

$$\delta = L_f - L_o$$

Axial Strain

$$\text{axial strain} = \frac{\text{change in length}}{\text{original length}}$$

$$e = \frac{\delta}{L_o}$$

Modulus of Elasticity

$$\text{modulus of elasticity} = \frac{\text{change in stress}}{\text{change in strain}}$$

$$E = \frac{\Delta s}{\Delta e}$$

Hooke's Law

$$\text{axial stress} = \text{modulous of elasticity} \times \text{axial strain}$$

$$s = Ee$$

Allowable Stress

$$\text{Allowable Stress} = \frac{\text{Yield Stress}}{\text{Factor of Safety}}$$

$$s_{allowable} = \frac{s_y}{n}$$

% Area of Reduction

$$\%\ \text{area reduction} = \frac{\text{original area} - \text{final area}}{\text{original area}}(100)$$

$$\%\ \text{area reduction} = \frac{A_0 - A_f}{A_o}(100)$$

% of Elongation

$$\%\ \text{elongation} = \frac{\text{final gage length} - \text{original gage length}}{\text{original gage length}}(100)$$

$$\%\ \text{elongation} = \frac{L_f - L_o}{L_o}(100)$$

Shear

$$\text{Shear Stress} = \frac{\text{Shear Force}}{\text{Shear Area}}$$

$$\tau = \frac{V}{A_s}$$

Torsion Testing

$$\tau_{\text{max shear stress at the outer surface}} = \frac{16T}{\pi d^3}$$

Area Moment of Inertia

$$I_{XX} = \frac{bh^3}{12}$$

Deflection Equations

$$\Delta_{\max} = \frac{FL^3}{48EI_{XX}}$$

$$s_{\text{flexure}} = \frac{3FL}{2bh^2}$$

Calculating Deflection Based on a Non-centered Force

$$\Delta_{\max} = \frac{Fa^2 b^2}{3EI_{XX} L}$$

Brinell Hardness Test

$$HB = \frac{2F}{\pi D (D - \sqrt{D^2 - d^2})}$$

Charpy Impact Test

$$E_{Charpy} = wr(\cos \beta - \cos \alpha)$$

Additional Reference Equations

Safety Factor

$$n = \frac{\text{actual stress limit}}{\text{allowable stress limit}}$$

Modulus of Resilience

$$\text{modulus of resilience} = \frac{1}{2}(\text{yield stress} \times \text{yield strain})$$

$$U_R = \frac{1}{2}(s_y e_y)$$

Modulus of Toughness

$$\text{modulus of toughness} \approx \frac{\text{yield stress} + \text{ultimate stress}}{2}(\text{fracture strain})$$

$$U_T \approx \frac{s_y 1 s_u}{2}(e_f)$$

For Brittle Metals

$$\text{modulus of toughness} \approx \frac{2}{3}(\text{ultimate stress} \times \text{fracture strain})$$

$$U_T \approx \frac{2}{3}(s_u e_f)$$

CHAPTER 11
The Manufacturing Process and Product Life-Cycle Management

Before You Begin

Think about these questions as you study the concepts in this chapter.

- How are raw materials converted into finished products?
- What are the methods for forming metal?
- How does metal forging work?
- How are extruded metal parts manufactured?
- What manufacturing processes are classified as cold forming?
- How are plastics processed?
- What manufacturing processes are ideal for making holes?
- What is meant by a product's *life cycle?*

BACKGROUND

Everything that you touch has had a manufacturing process performed on it. Knowing the different ways that objects are made will enhance your ability to create the highest-quality products in the most time- and cost-efficient method. This unit will look at a variety of processes for metallic and nonmetallic materials.

Casting Processes: Sand, Investment, Die, Permanent Mold

There are a variety of ways to mold and cast materials to make them into objects. When you bake a cake, you are using a mold to form the cake. The Society of Manufacturing Engineers has created a series of videos to help understand these different processes. A general collection of videos can be found at the following website:

http://www.sme.org/fmp/?terms=casting%20process%20video

Other SME videos are found at the following YouTube links:

Die Casting http://www.sme.org/ProductDetail.aspx?id=22200

Sand Casting http://www.youtube.com/watch?v=mx1qteRUYwI

SME Thread Rolling http://www.youtube.com/watch?v=omRsmuNp8Ec

Exercise 11.1 Metal Product Process Identification

Objective

Identify the process that was used to create various objects.

Procedure

1. Select three different metal household products that have been casted.
2. Document the items by photographing them and identify the type of casting method used.

Exercise 11.2 Mold Casting Project for Home or School

Objective

Apply a mold casting process.

Materials

- Microwave-safe bowl
- Semisweet chocolate chips (made by two different manufacturers)
- Candy thermometer or instant-read thermometer
- Silicone ice cube tray or candymaking silicone mold
- Spatula
- Paper towels for cleanup

Procedure

The process of casting involves heating a material past its melting point so that it flows easily into a mold. To do this, perform the following steps:

1. Place some of the semisweet chocolate in a microwave-safe bowl and heat the chocolate so it melts into a thick liquid. Measure the temperature of the chocolate just as it begins to melt. Take the bowl out of the microwave before you check the temperature (be careful—the bowl could be hot, even if it's microwave-safe!).

2. Pour the chocolate into a silicone ice cube tray or candymaking silicone mold. It is more difficult to manage the flowing chocolate than you might think, so pour slowly.
3. Let the chocolate sit at room temperature until it loses enough heat to solidify. This process can be sped up by placing into a refrigerator or freezer. To see the effects of temperature change on chocolate fully, do one instance of each method and see what the different results are from cooling the chocolate with different methods.
4. Clean up the mess while you are waiting for the chocolate to harden.

Conclusion

1. How much time and how much energy were used to melt the chocolate? Use the candy thermometer to measure the temperature. Record the wattage of the microwave as well as the time it took to do this exercise.
2. Chart the temperature of each student's reading. What is the cause of the variability in the readings?
3. How long did it take for the chocolate to cool at room temperature? In a refrigerator or freezer?
4. If the chocolate is placed into the refrigerator or freezer, did the surface, consistency, or structure of the chocolate change? Is the taste any different?
5. Note which chocolate is harder after the cooling process.

Exercise 11.3 Golf Club Heads

Objective

This exercise will help identify the processes of how golf club heads are made, either through casting or through forging. Many golf club heads are investment-casted using a ceramic shell over a wax-based mold that is then melted away. The very expensive club heads are drop-forged rather than cast. Drop-forging uses a multistep die and presses the metal into shape.

Procedure

Stop at a local golf course or golf supply store and ask to see different manufacturers' golf clubs and putters. See if you can determine by sight and touch if the club head has been casted or forged. Many times, this information is in the literature or can be found online on a manufacturer's website.

Conclusion

Identify three different clubs that are cast and three that are forged. Take pictures of the clubs and write a short report linking the cost basis of the club heads to whether the clubs were forged or cast.

Exercise 11.4 Melting Points

Objective

Identify the melting point of common metals used in the casting process.

Procedure

Research the melting points of the following metals and record them in Table 11-1.

Melting Point Table	
Metal	**Melting Point (°F or °C)**
Aluminum	
Pot metal (a combination of several types of metals)	
Bronze	
Nickel	
Silver	
Copper	
Iron	

TABLE 11-1 *Melting Point.*

Metal Forming

When metal is formed, heat and pressure are used to change the shape of the metal into an object.

Forging, Rolling, Extrusion, Deep Drawing, and Metal Spinning

There are many different methods for forming metals, including forging, rolling, extrusion, deep drawing, and spinning. Here are some links to videos that illustrate various metal-forming processes:

SME Video on Sheet Metal Coil Processing http://www.sme.org/ProductDetail.aspx?id=22207

SME Video on Forging http://www.sme.org/ProductDetail.aspx?id=22245

Exercise 11.5 Play-Doh™ or Cookie Dough Stamping

Objective

Use a stamping process to form an object.

Materials

- Play-Doh or a batch of premade cookie dough
- A set of cookie cutters that will leave a pattern on top of the cookie (that is, not open cutters)

Procedure

1. Roll out the Play-Doh or cookie dough so that it is thicker than the height of the cookie cutter you are using.
2. Press the cookie cutter into the dough to cut, imprint, and form a pattern.
3. Repeat for the complete rolled-out sheet of dough.
4. Remove the excess material that is left by the cookie cutter so it can be reused.

Conclusion

1. How close can the cookie cutter patterns be placed?
2. Does increasing the pressure on the cutter produce better cutting results?

3. Does the type and consistency of the dough make a difference in the quality of the formed patterns?
4. Weigh the Play-Doh or cookie dough before you begin, and then weigh the excess that did not get used at the end of the exercise. What is the difference?
5. Can the remainder get reused to make additional pieces? If so, how many more pieces can be created?
6. What is the weight of the dough after the next set of cuts are made?
7. At what point does the time and effort of the labor outweigh the benefit of making another mold cut?

Plastics

Plastics are used to make many of the items you use today. There are a large variety of different plastic materials, as well as corresponding processes used on these materials to create products. This section is on different processes that are used on plastics to create objects.

Blow Molding, Thermoforming, Extrusion, and Injection Molding

Like metals, plastics can be formed by different methods, including blow molding, thermoforming, extrusion, and injection molding. Here is a list of links to videos that illustrate various processes to create plastic products.

SME Video on Thermoforming http://www.sme.org/ProductDetail.aspx?id=22322

SME Video on Blow Molding http://www.sme.org/ProductDetail.aspx?id=22198

SME Video on Injection Molding http://www.sme.org/ProductDetail.aspx?id=22263http://www.sme.org/ProductDetail.aspx?id=22199

Exercise 11.6 Thermoform a Fruit Roll-Up in an Aluminum Bowl

Objective

Apply thermoforming techniques to create an object.

Materials

- A fruit roll-up style snack
- Small aluminum bowl (plastic will work, but aluminum is better)
- Ice
- Larger bowl for the ice water bath
- Hair dryer (to warm the fruit roll-up style snack)

Procedure

1. Unroll and place the fruit roll-up on the surface of the small aluminum bowl, either on the inside or on the outside (if more than one fruit roll-up is used, make sure that the seam or seams overlap).
2. Warm the fruit roll-up with the hair dryer on low heat and air speed. (We do not want to melt the roll up; we want it to warm up so that it is flexible.)
3. Once the fruit roll-up has conformed to the bowl, then stop the heat and unplug the hair dryer for safety.
4. Place the aluminum bowl in the bowl of ice water to cool the roll-up, which will slightly shrink and release it from the bowl. (Do not let the fruit roll-up touch the water.)
5. The goal of thermoforming is to hold the shape of the object that the item formed. Once the fruit roll-up has cooled and solidified, it should maintain the shape of the bowl.

Conclusion

1. Did the fruit roll-up conform to the bowl? Did it maintain its shape after cooling?
2. If the fruit roll-up did not maintain the shape of the bowl, what could you do to get different results next time?
3. If too much heat is placed on the fruit roll-up, what happens to the roll-up?
4. What other types of food materials can be formed using a combination of heat and pressure?
5. What other nonfood materials could be used for this type of experiment? (Materials that are soft but maintain their shape when cooled?)

Chip-Producing Processes

Chip-producing methods produce finished parts by cutting material into chips. They include sawing, drilling, turning, and milling.

Sawing, Drilling, Turning, and Milling

Here is a list of links to videos that illustrate various chip-producing processes.

SME Video on Hole Drilling http://www.sme.org/ProductDetail.aspx?id=22345

SME Video on Turning and Lathe Work http://www.youtube.com/watch?v=tDc0l9Gm8D4&feature=mfu_in_order&list=UL

SME Video on a Milling and Machining Center http://www.youtube.com/watch?v=SeGQ91FKTNg&feature=relmfu

Exercise 11.7 Drill Sizing

Objective

Apply drilling techniques and quality standards to determine the differences in a process that contribute to a change in diameter.

Materials

- Drill
- Drill bit (1/4" or larger)
- 2" × 4" scrap piece
- Calipers
- Recording sheet

Procedure

1. Lay out the drilling positions and drill five holes into a board using the same bit.
2. Measure the holes with a set of calipers.

Conclusion

1. Why are the holes different sizes? Determine what the variability may be in the process that you used and how that variability can be minimized.
2. How can the variability of the hole sizes be minimized?
3. Can factors such as heat of the drill bit and air humidity influence the hole size?

Product Life-Cycle Management (PLM)

A product life cycle begins when a product is first introduced into the market and ends when that product is taken off the market.

PLM Process

The process of product life-cycle management (PLM) shows a product from the cycle of birth to disposal. Being able to document this process up-front, before a product is developed, will foster the overall understanding of how the product processes and support will need to be applied throughout the product's life. The goal of PLM is to track and share the data that is relevant to the processes among all the major partners in a manufacturing business. This includes everyone from the CEO to the person who cleans the machines.

This data is usually stored on a server that is accessible by team members both internal to the company and suppliers outside the company. This exchange of data is termed product data management (PDM).

Exercise 11.8 PLM Process Diagrams

Objective

Identify different PLM process cycles that are used in industry.

Procedure

Your textbook illustrates one example of a PLM process diagram (see Figure 11-67). There are many different process diagrams that vary in detail for PLM.

1. Research and find two different PLM process charts that are used in industry. (Hint: look for images that reflect a PLM process.)
2. Printing the charts, or otherwise putting them side by side, develop a common thread matrix between them. (What steps match up between the PLM process charts?)
3. In a matrix format, show how the steps match up between each of the PLM models. An example matrix chart matching up three different PLM processes is illustrated in Figure 11-1. The arrows show the common links between the processes.

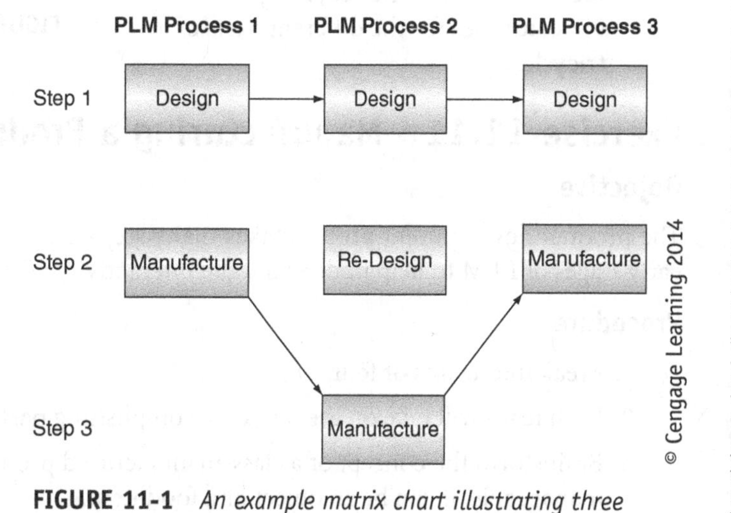

FIGURE 11-1 *An example matrix chart illustrating three different PLM processes.*

Exercise 11.9 PLM Process Identification for Products

Objective

Apply the PLM process by having students develop the process using a product as an example.

Procedure

Using an item from home or in the classroom, try to fill out the PLM circle as completely as possible. Use the Internet and company research to assist in finding components of PLM.

Use the generic chart illustrated in Figure 11-2 as a guide to specify how the product will interact at each step. A short statement about each step in the PLM chart is the deliverable.

Exercise 11.10 Team-Based Maintenance, Service, Product Disposal, and Recycling

Objective

One of the most important but often overlooked areas in PLM is the last two items (Maintenance and Repair and Disposal/Recycling). Using the example provided, create a maintenance and service manual and identify disposal and recycling opportunities for the product.

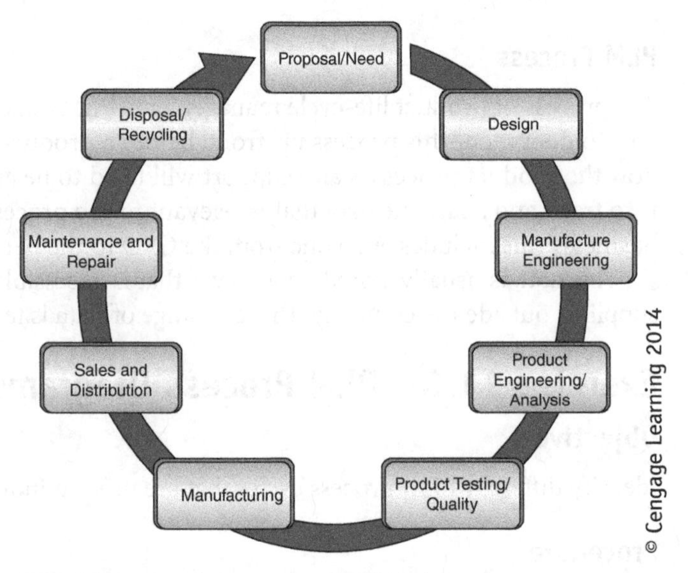

FIGURE 11-2 *An example of a PLM chart.*

Procedure

Using a child's tricycle as an example, perform the following steps. The deliverable will be a maintenance and service manual of the tricycle, the last page of which will address the disposal and recycling of the tricycle parts.

1. Write a maintenance and service manual.
2. Identify disposal and recycling opportunities for the different parts of a tricycle.

Exercise 11.11 Manufacturing a Product

Objective

The product development process takes many steps. Apply the product development cycle that is used in the early stages of PLM to help understand the interactivity of the PLM process.

Procedure

1. Break into teams of four.
2. Each team will take on the task of accomplishing part of the PLM process.
3. Brainstorm the concept of a class-manufactured product. Keep it simple, with few parts, so it stays manageable. It can be an object or a food item.
4. Once a product is selected, each team will need to choose a part of the PLM process:
   a. Administrative team—takes on total responsibility of project; communicates with all teams.
   b. Design team—responsible for product engineering.
   c. Supply chain—responsible for acquisitions.
   d. Manufacturing (two teams)—one team works on the manufacturing design and the other works as a production team.
   e. Marketing/sales—responsible for marketing the product.
   f. Quality control of product—handles quality control; reports to the Manufacturing Design team.
   g. Service-repair-disposal—responsible for servicing and disposing of the product; reports to the Manufacturing and Marketing/sales teams.
5. Work toward actually making a product in the classroom.

Conclusion

1. Identify areas of breakdown in the system that you developed.
2. What worked very well with your product process, and how well integrated was the PLM process with the product?
3. How would you have run this exercise differently if you were in charge?

CHAPTER 12
Statics

Before You Begin

Think about these questions as you study the concepts in this chapter.

- What do civil engineers do?
- What is the difference between a scalar and a vector quantity?
- What does it mean if an object or structure is in static equilibrium?
- What are free-body diagrams, and when are they used?
- How do you determine the direction and magnitude of a reaction force that occurs at a beam or truss support location?
- How do you determine the magnitude and type of force that is transmitted through a truss member?

BACKGROUND

Careers in Engineering

Explore Your World

Professional societies can be an excellent source of information about engineering disciplines. Founded in 1852, the American Society of Civil Engineers (ASCE) represents more than 144,000 civil engineers worldwide, and, by its own account, it is America's oldest national engineering society. Visit the ACSE website (http://www.asce.org/) to find professional publications, learn about career opportunities in civil engineering, download West Point Bridge Designer software, locate local ASCE chapters, and much more.

Consider contacting a civil engineering professional in your community to:

- Request that he or she be your mentor on an engineering design project
- Invite her or him to visit your classroom and talk about the engineering profession
- Arrange a job shadowing opportunity at his or her workplace

Reconciling Vectors

Exercise 12.1 Adding Vectors

Objective

At the conclusion of this exercise, you will be able to do the following:

1. Reconcile vectors using the parallelogram method of vector addition
2. Reconcile vectors using the tip-to-tail method of vector addition
3. Reconcile vectors using the component method of vector addition

Procedure

Read the section on vector addition (pp. 403–406) and the section on free-body diagrams (p. 409) in Chapter 12, "Statics," of your *Principles of Engineering* textbook.

Materials

- Protractor
- Standard inch rule, with an accuracy of 1/8 inch (see Figure 14-4 on page 449 of your textbook)

Problem 12.1 Figure 12-1 is a scaled vector drawing representing two forces, F_1 and F_2, that act on an object.

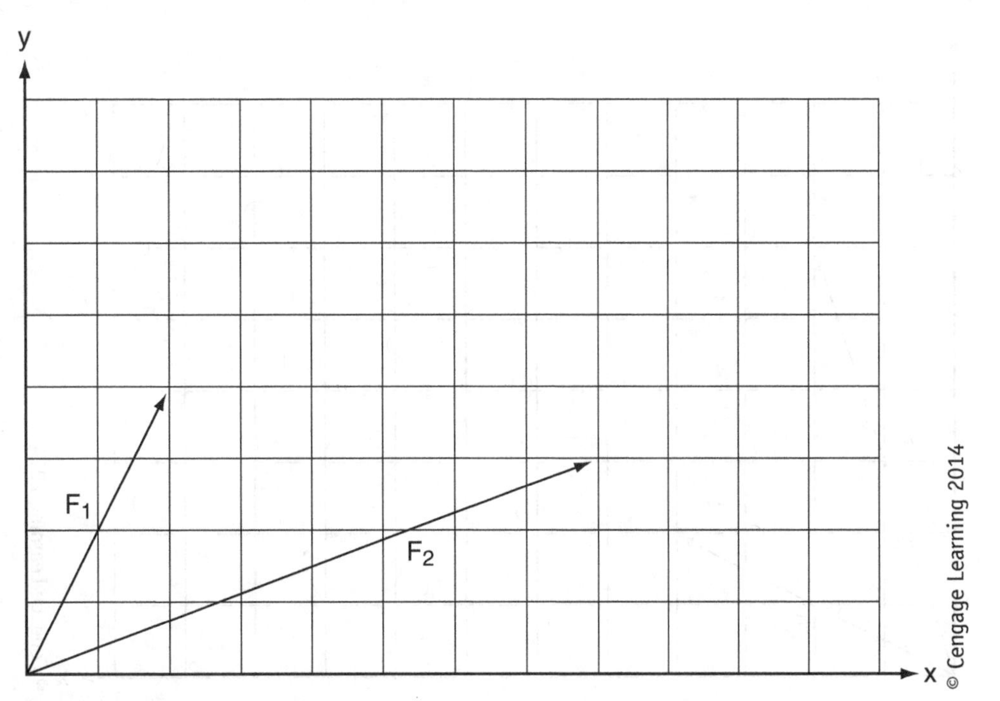

FIGURE 12-1 *Vector addition—parallelogram method.*

1. Draw the resultant force vector (F_R) on Figure 12-1 using the *parallelogram method* of vector addition, described on page 404 of your textbook.
2. Measure the length of resultant force vector (FR) to the nearest 1/8-inch and record its length in the space provided.
   Length of F_R = ______________in.
3. Determine the magnitude of the resultant force (F_R) using the scale (1/8 in. = 1 lb) and record the value in the space provided.
   F_R = ______________lb
4. Measure the angle of resultant force vector (F_R) relative to the horizontal axis and record its direction in the space provided.
   θ = ______________°

Problem 12.2 Figure 12-2 is a scaled vector drawing representing two forces, F_1 and F_2, that act on an object.

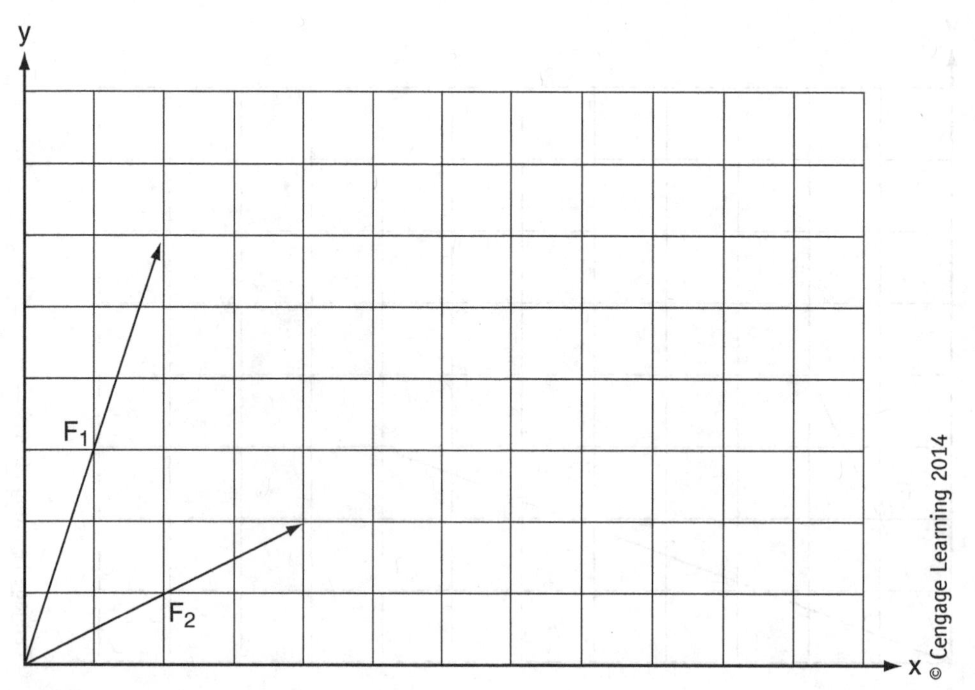

FIGURE 12-2 *Vector addition—parallelogram method.*

1. Draw the resultant force vector (FR) on Figure 12-2 using the *parallelogram method* of vector addition described on page 404 of your textbook.
2. Measure the length of resultant force vector (F_R) to the nearest 1/8-inch and record its length in the space provided.
   Length of F_R = ______________ in.
3. Determine the magnitude of the resultant force (F_R) using the scale 1/8 in. = 1 lb and record the value in the space provided.
   F_R = ______________ lb
4. Measure the angle of resultant force vector (FR) relative to the horizontal axis and record its direction in the space provided.
   θ = ______________ °

Problem 12.3 Figure 12-3a is a scaled vector drawing representing three forces, F_1, F_2, and F_3, that all act on an object.

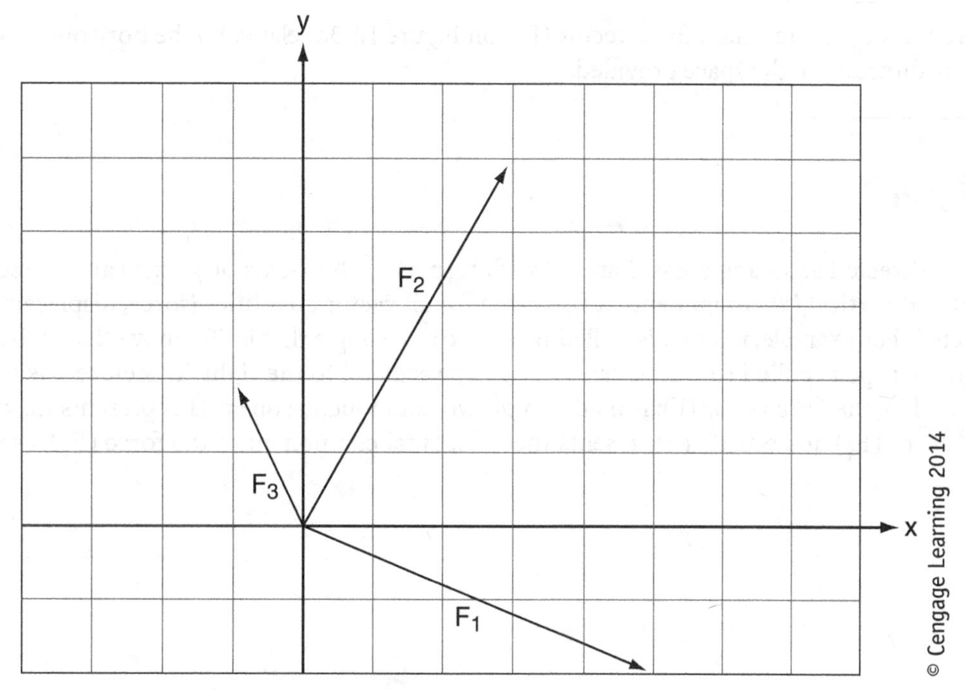

FIGURE 12-3a *Vector addition—tip-to-tail method.*

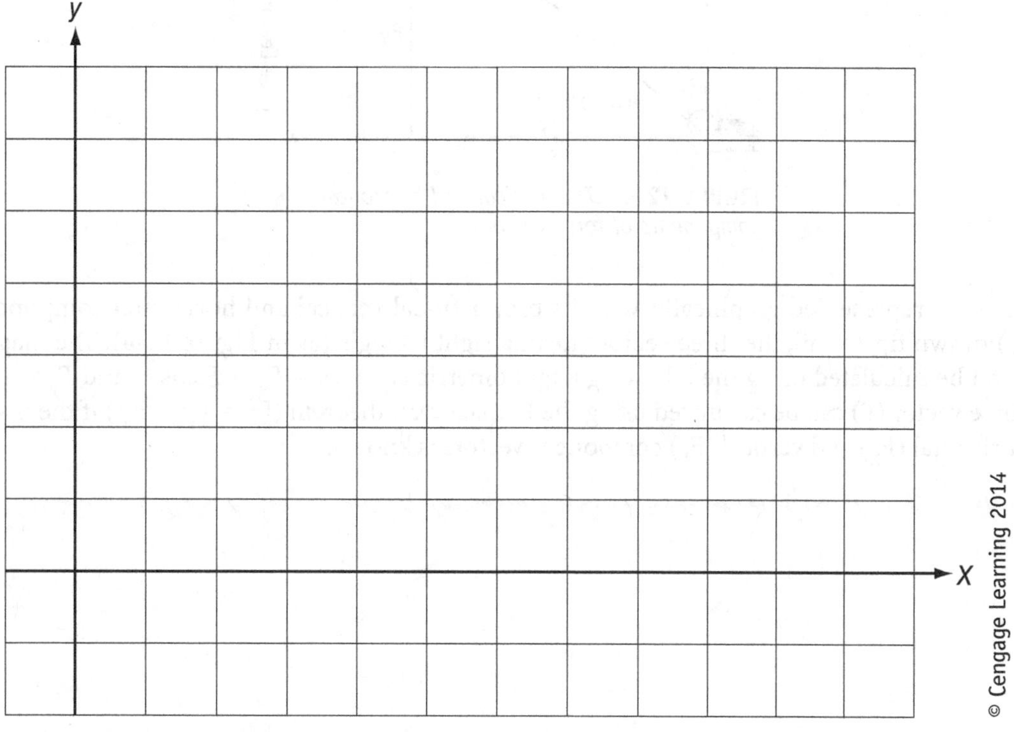

FIGURE 12-3b *Using the tip-to-tail method to draw vectors.*

1. Redraw force vectors F_1, F_2, and F_3 on Figure 12-3b using the *tip-to-tail* method of vector addition described in your textbook. Then draw the resultant force vector (F_R) to complete the exercise.
2. Measure the length of resultant force vector (F_R) on Figure 12-3b to the nearest 1/8-inch and record its length in the space provided.
   Length of F_R = ______________in.

3. Determine the magnitude of the resultant force (F_R) using the scale 1/8 in. = 1 lb and record the value in the space provided.
   F_R = ______________lb
4. Measure the angle of resultant force vector (F_R) on Figure 12-3a relative to the horizontal axis and record its direction in the space provided.
   θ = ______________°

TIP SHEET

A vector that is directed at an angle less than 90° with respect to the *x*-axis or *y*-axis can be resolved into its horizontal (*x*) and vertical (*y*) components. This is true for all vector quantities (force, displacement, velocity, acceleration, etc.). For example, if a sled is pulled by a rope at an angle that is 40° above the horizontal (*x*) axis (Figure 12-4), the rope is pulled (or has tension) both upward and to the right. Therefore, the tension in the rope, represented by the *force vector* (F), is made up of two components: one that represents the vertical component of the force (F_Y) and one that represents the horizontal component of the force (F_X), as illustrated in Figure 12-4.

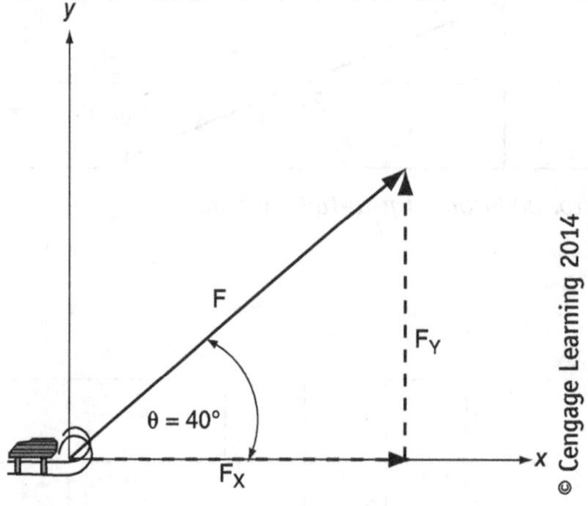

FIGURE 12-4 *The vertical and horizontal components of force vector F.*

When a vector is represented graphically, with its proportional vertical and horizontal component vectors (F_X and F_Y) drawn tip-to-tail, the three vectors form a right triangle (as in Figure 12-4). The magnitudes of F_X and F_Y can be calculated using the following trigonometric equations: $F_x = F \cos \theta$, and $F_y = F \sin \theta$. The resultant force vector (F) can be calculated using the Pythagorean theorem ($F^2 = F_x^2 + F_y^2$) if the magnitude of both the horizontal (F_X) and vertical (F_Y) component vectors is known.

Problem 12.4 Figure 12-5 shows the vector representation of resultant force (F) and its vertical (F_Y) and horizontal (F_X) component vectors.

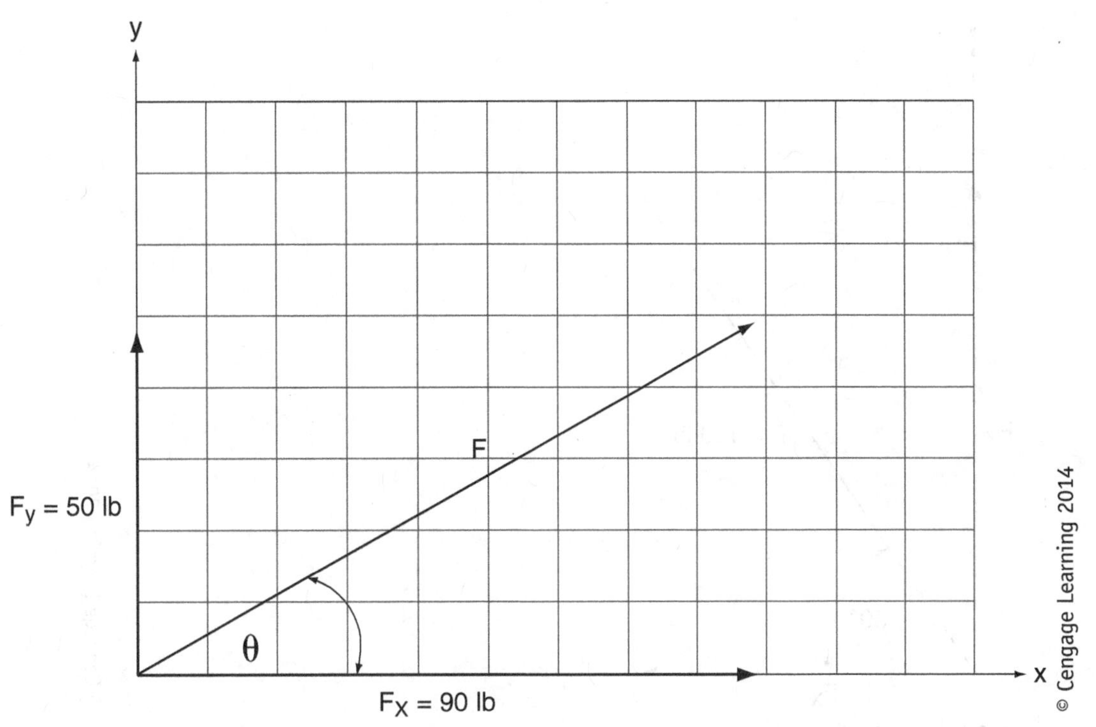

FIGURE 12-5 *Vector resolution—component method.*

1. Redraw F_y so that its tail begins at the tip of F_x and its tip touches the tip of F. Label this force F_y.
2. Calculate the magnitude of force (F) shown in Figure 12-4 using the Pythagorean theorem. Show your math work and record your answer in the space provided.

F = ________________ lb

3. Calculate the angle (θ) of force vector (F). Show your math work and record your answer in the space provided.

θ = ________________

Problem 12.5 Figure 12-6 is the vector representation of a 120-lb force (F) exerted at an angle of 50° relative to the horizontal axis and the horizontal (F_X) and vertical (F_Y) components of force F.

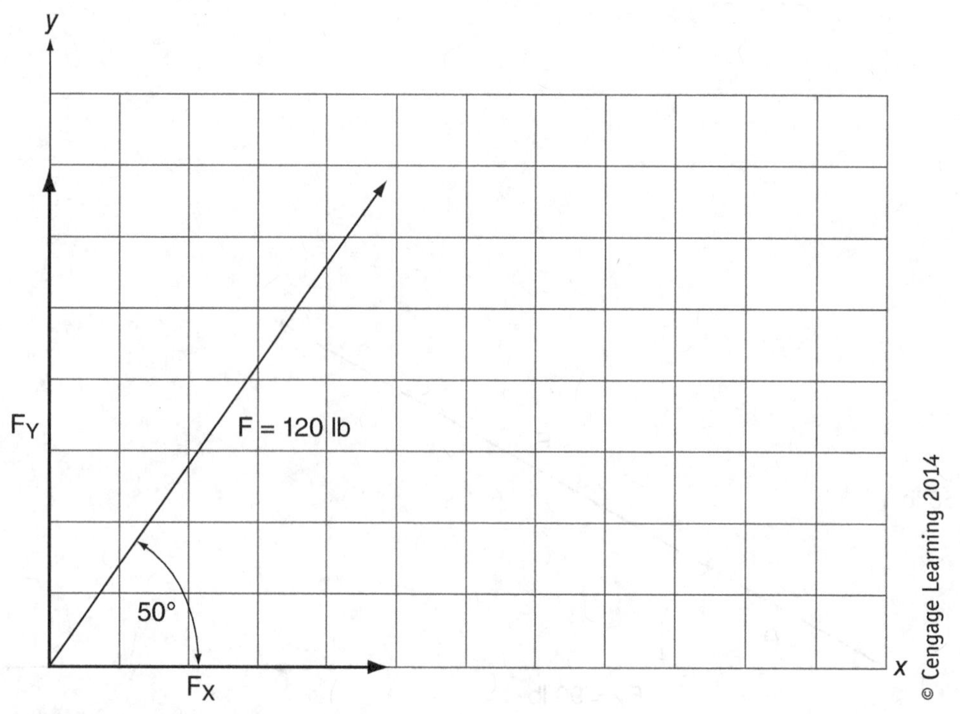

FIGURE 12-6 *Vector resolution—component method.*

1. Redraw F_y so that its tail begins at the tip of F_x and its tip touches the tip of F. Label this force F_y.
2. Calculate the magnitude of the horizontal component force (F_X). Show your math work and record your answer in the space provided.

F_X = ________________lb

3. Calculate the magnitude of the vertical component force (F_Y). Show your math work and record your answer in the space provided.

F_Y = ________________lb

TIP SHEET

Vectors are often represented on a two-dimensional Cartesian grid with their tails at the intersection where $x = 0$ and $y = 0$. This has already been shown in many figures in this chapter, including Figure 12-6. In all these instances, however, all vectors were shown in Quadrant I (Figure 12-7). In Quadrant I, vectors (again, with their tails at $x = 0$ and $y = 0$), point up and to the right, having positive x and y values. However, if a vector were in Quadrant II, it would still point upward, having a positive y value, but it would point to the left, indicating a negative x value. If a vector were in Quadrant III, pointing down and to the left, both the x and y values would be negative. Finally, if a vector were in Quadrant IV, pointing down and to the right, its y value would be negative and its x value would be positive.

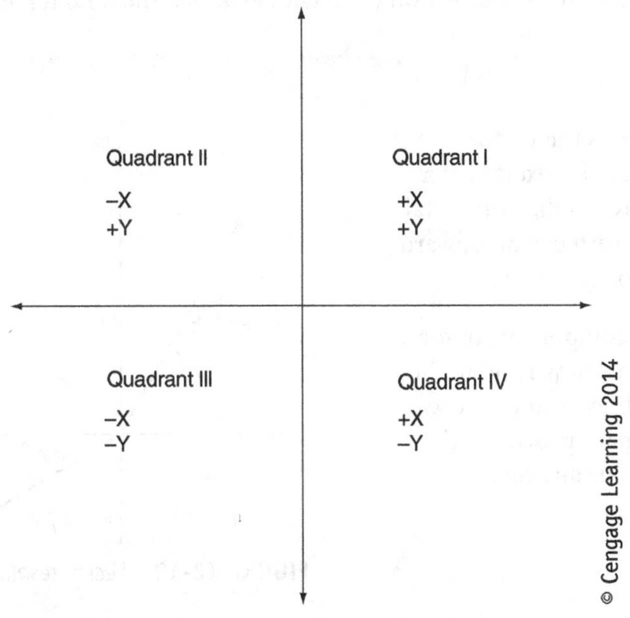

FIGURE 12-7 *The four quadrants of the Cartesian coordinate axes.*

The Cartesian grid provides a "sign convention" for assigning positive and negative values to vectors that are parallel to the x or y axis. A vector that is parallel to the x axis (e.g., F_x) that points to the right is given a positive sign, and one that points to the left is given a negative sign. Likewise, a vector that is parallel to the y axis (e.g., F_y) that points upward is given a positive sign, and one that points downward is given a negative sign. When adding component vectors together, their signs must be considered (Figures 12-8 and 12-9).

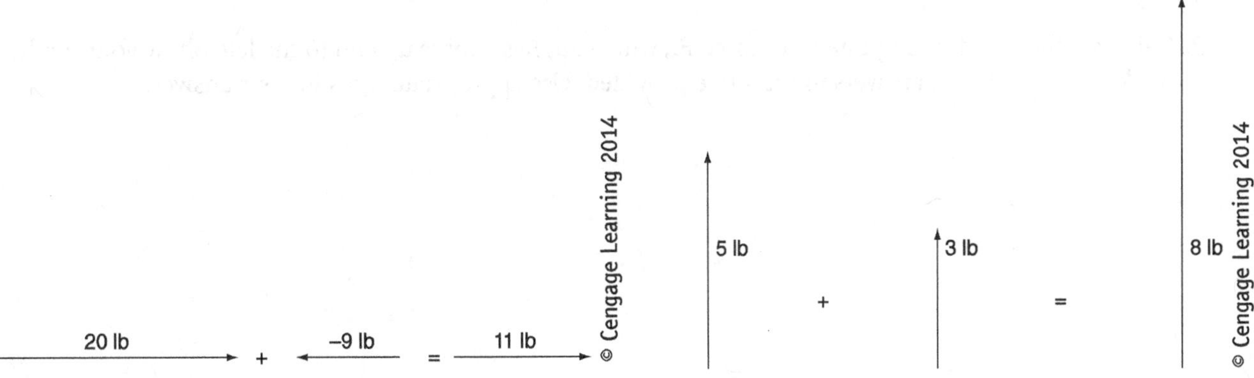

FIGURE 12-8 *Adding horizontal component vectors.*

FIGURE 12-9 *Adding vertical component vectors.*

When multiple forces act upon an object at various angles, a good strategy to determine the resultant force acting on the object (in other words to add those vectors together) is to do the following:

1. Use trigonometry to break each vector down into component parts (F_x and F_y).
2. Use the sign convention described above to add all the forces in the *x* direction together to get the resultant force acting on the object in the *x* direction, F_{RX}.
3. Use the sign convention to add all the forces in the *y* direction together to get the resultant force acting on the object in the *y* direction, F_{RY}.
4. Use the Pythagorean theorem to determine the magnitude of the total force, F_R.
5. Use trigonometry to determine the direction (or angle) at which the total force acts.

Problem 12.6 Figure 12-10 is the vector representation of a 58-lb force (F_1) exerted at a downward angle of 30° relative to the horizontal axis and an 80-lb force (F_2) exerted at an upward angle of 60° relative to the horizontal axis.

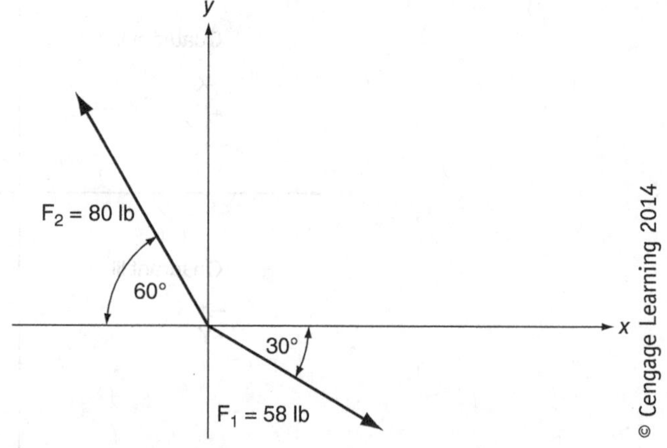

FIGURE 12-10 *Vector resolution—component method.*

1. Calculate the *x* and *y* components of force F_1, which applies a force down and to the right. Show your math work and record your answers in the space provided. Use appropriate signs in your answers.

F_{1X} = ____________lb F_{1Y} = ____________lb

2. Calculate the *x* and *y* components of force F_2, which applies a force up and to the left. Show your math work and record your answers in the space provided. Use appropriate signs in your answers.

F_{2X} = ____________lb F_{2Y} = ____________lb

3. Calculate the resultant force exerted in the x direction, F_{RX}, and in the y direction, F_{RY}. Show your math work and record your answers in the space provided. Use appropriate signs in your answers.

F_{RX} = ____________lb F_{RY} = ____________lb

4. Calculate the magnitude and direction of the resultant force, F_R. Show your math work and record your answers in the space provided.

F_R = ____________lb θ = ____________°

5. Using the values from Problem 12.5(3), sketch and label the resultant force vector F_R on Figure 12-10.

Problem 12.7 Figure 12-11 shows an eye hook assembly that is attached to a load and is supported by two separate cables. F_1 and F_2 represent the tension force in the cables of 150 lb and 80 lb, respectively.

6. Use the space provided to draw F_1 and F_2 on a Cartesian grid.

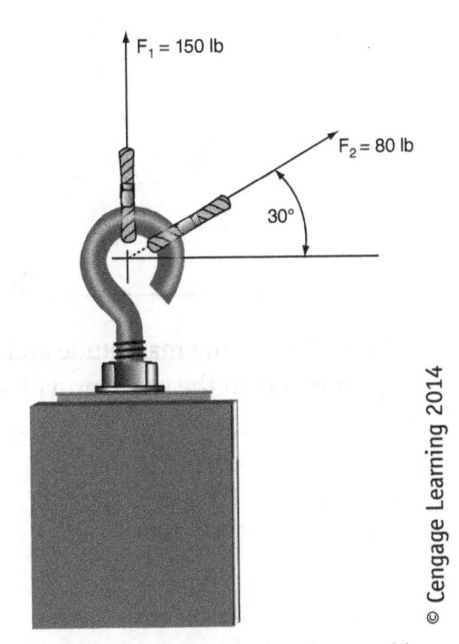

FIGURE 12-11 *Eye hook assembly.*

7. What are the x and y components of force F_1? Record your answers in the space provided. Use appropriate signs in your answers.

F_{1X} = ____________lb F_{1Y} = ____________lb

8. Calculate the x and y components of force F_2. Show your math work and record your answers in the space provided. Use appropriate signs in your answers.

F_{2X} = ____________lb F_{2Y} = ____________lb

9. Calculate the resultant force that the cables exert on the eye hook in the x direction, F_{RX}, and in the y direction, F_{RY}. Show your math work and record your answers in the space provided. Use appropriate signs in your answers.

F_{RX} = ____________lb F_{RY} = ____________lb

10. Calculate the magnitude and direction of the resultant force. Show your math work and record your answers in the space provided.

F_R = ____________lb θ = ____________°

Exercise 12.2 Calculating Support Reactions

Objective

At the conclusion of this exercise, you will be able to do the following:

1. Identify a pinned support and a roller support.
2. Draw a free-body diagram of a simply supported beam.
3. Apply the equations of equilibrium to calculate the support reactions on a beam.

Procedure

Read the section on simple supports and reactions (pp. 409–413) in Chapter 12 of your textbook.

TIP SHEET

In a simply supported beam, a pinned support ( ) generates both a vertical (y-direction) and a horizontal (x-direction) reaction force.

A roller support ( ) generates only a vertical (y-direction) reaction force.
When applying the equilibrium equation $\Sigma M = 0$ (the sum of the moments about a point is equal to zero) to calculate the support reactions on a beam or truss, assume that all forces that cause clockwise rotation ↻ about a point are negative and all forces that cause counterclockwise ↺ rotation about a point are positive.

Two other equilibrium equations will be helpful as you practice these exercises: $\Sigma F_x = 0$ (sum of forces in the x direction = 0) and $\Sigma F_y = 0$ (sum of forces in the y direction = 0). These remind us that the beams are not accelerating—indeed, they are at rest.

Problem 12.8 Figure 12-12 shows a simply supported beam with a concentrated load of 125 lb applied at point B.

1. In the space provided, draw and label a free-body diagram of the beam:

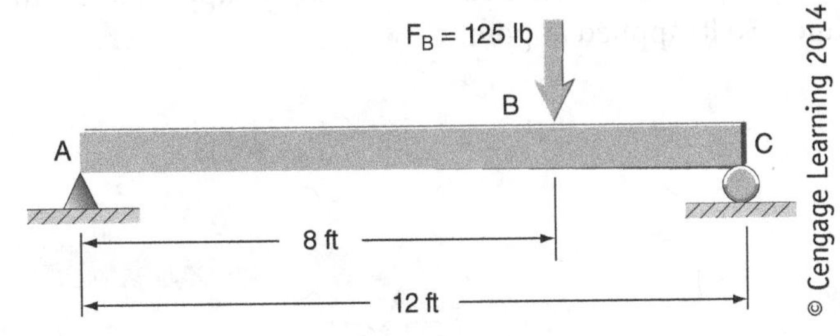

FIGURE 12-12 *A simply supported beam.*

2. Calculate the vertical reaction force at point C. Show your math work and record your answer in the space provided. Use the appropriate sign in your answer.

R_{CY} = _______________ lb

3. Calculate the vertical reaction force at point A. Show your math work and record your answer in the space provided. Use the appropriate sign in your answer.

R_{AY} = _______________ lb

4. What is R_{AX}? How do you know?

____

____

Problem 12.9 Figure 12-13 is a simply supported beam with concentrated loads of 500 lb applied at point B and 150 lb applied at point C.

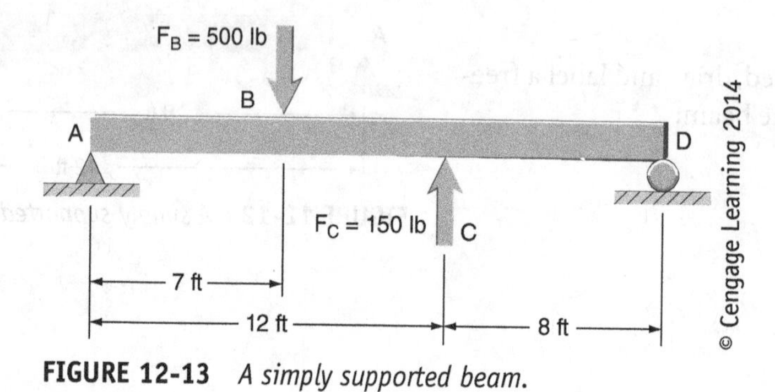

FIGURE 12-13 *A simply supported beam.*

1. In the space provided, draw and label a free-body diagram of the beam:

2. Calculate the vertical reaction force at point D. Show your math work and record your answer in the space provided. Use the appropriate sign in your answer.

R_{DY} = ________________lb

3. Calculate the vertical reaction force at point A. Show your math work and record your answer in the space provided. Use the appropriate sign in your answer.

R_{AY} = ________________lb

4. What is R_{AX}? How do you know?

____

____

Problem 12.10 Figure 12-14 is a simply supported beam that has a concentrated 800-lb load applied at a 30° angle at point B and a concentrated 200-lb load applied at point C.

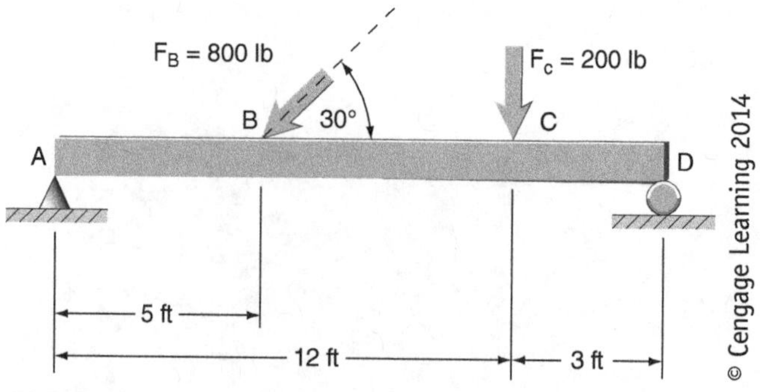

FIGURE 12-14 *A simply supported beam.*

1. In the space provided, draw and label a free-body diagram of the beam:

2. Calculate the vertical and horizontal components of force F_B. Show your math work and record your answers in the space provided. Use the appropriate signs in your answer.

F_{BX} = ______________________ lb F_{BY} = ______________________ lb

3. Calculate the vertical reaction force at point D. Show your math work and record your answer in the space provided. Use the appropriate sign in your answer.

R_{DY} = _______________ lb

4. Calculate the vertical reaction force at point A. Show your math work and record your answer in the space provided. Use the appropriate sign in your answer.

R_{AY} = _______________ lb

5. Calculate the horizontal reaction force at point A. Show your math work and record your answer in the space provided. Use the appropriate sign in your answer.

R_{AX} = _______________ lb

Structural Analysis of Trusses

Exercise 12.3 Calculating Reactions and Internal Forces of Truss Systems

Objective

At the conclusion of this exercise, you will be able to do the following:

1. Check for static determinacy of a truss.
2. Draw a free-body diagram of a truss system.
3. Apply the equations of equilibrium to determine the internal member forces in a truss system.

Procedure

Read the section on the structural analysis of trusses (pp. 414–423) in Chapter 12 of your textbook.

Problem 12.11 Figure 12-15 illustrates four truss systems, labeled A, B, C, and D.

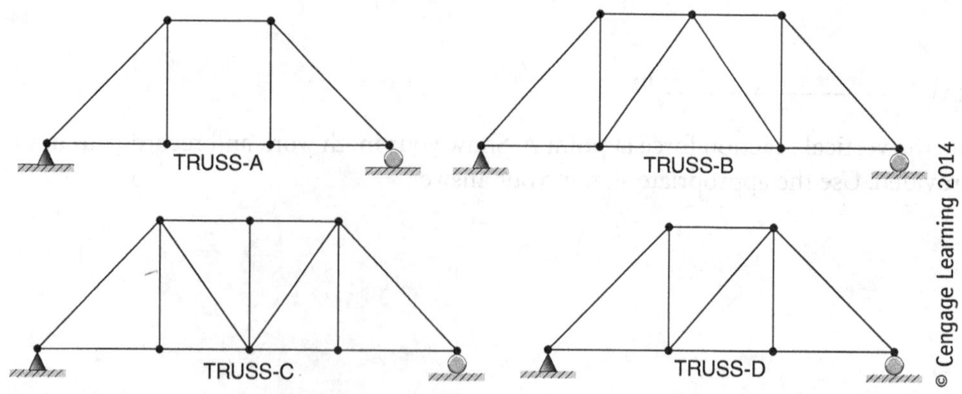

FIGURE 12-15 *Static determinacy of truss systems.*

Using the method described on page 415 of your textbook, analyze each of the truss systems to determine if it is a *statically determinate* truss. Show your math work for each truss and indicate your conclusion in the space provided.

TRUSS A—Statically determinate? Yes________ No__________

TRUSS B—Statically determinate? Yes________ No__________

TRUSS C—Statically determinate? Yes________ No__________

TRUSS D—Statically determinate? Yes________ No__________

Problem 12.12 Figure 12-16 is a statically determinate truss that has one pinned support at point A, one roller support at point C and a concentrated load applied at point B.

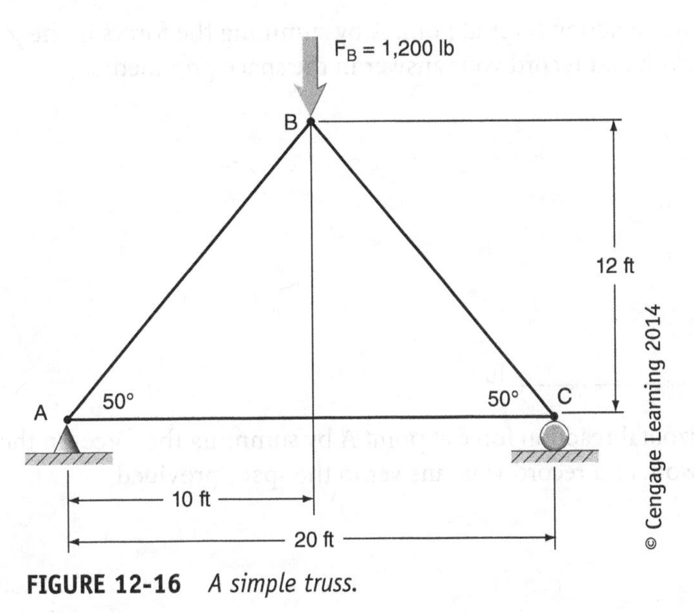

FIGURE 12-16 *A simple truss.*

1. In the space provided, draw and label a free-body diagram of the entire truss. *For reference, see Figure 12-43 on page 417 of your textbook.*

2. Calculate the vertical reaction force at point C by summing the moments about point A ($\Sigma M_A = 0$). Show your math work and record your answer in the space provided.

R_{CY} = ____________________lb

3. Calculate the vertical reaction force at point A by summing the forces in the *y*-direction ($\Sigma F_Y = 0$). Show your math work and record your answer in the space provided.

R_{AY} = ____________________lb

4. Calculate the horizontal reaction force at point A by summing the forces in the *x*-direction ($\Sigma F_X = 0$). Show your math work and record your answer in the space provided.

R_{AX} = ____________________lb

5. In the space provided, draw and label a free-body diagram of each joint, showing all forces acting on and within the truss. *For consistency, assume all members to be in tension and draw force arrows (F_{AB}, F_{AC}, and F_{BC}) pointing away from the joints. For reference, see Figure 12-44 on page 419 of your textbook.*

6. Determine the number of unknown forces at each joint and complete Table 12-1.

Joint	Equilibrium Equation	# Unknowns
A	$\Sigma F_X = 0$	________
A	$\Sigma F_Y = 0$	________
B	$\Sigma F_X = 0$	________
B	$\Sigma F_Y = 0$	________
C	$\Sigma F_X = 0$	________
C	$\Sigma F_Y = 0$	________

TABLE 12.1 *Unknown Forces.*

For Problems 12.12 through 12.14, show your math work in the space provided after each problem. Record your answers in Table 12-2 and indicate whether the member is in tension or compression.

Problem 12.13 Calculate the internal member forces for truss section AB using equilibrium equation $\Sigma F_{AY} = 0$.

Problem 12.14 Calculate the internal member forces for truss section AC using equilibrium equation $\Sigma F_{AX} = 0$.

Problem 12.15 Calculate the internal member forces for truss section BC using equilibrium equation $\Sigma F_{CY} = 0$.

Truss Member	Internal Force (lb)	Tension/Compression
AB	F_{AB} = ____________	____________
AC	F_{AC} = ____________	____________
BC	F_{BC} = ____________	____________

TABLE 12.2 *Truss Member Internal Forces.*

Problem 12.16 Figure 12-17 is a statically determinate truss that has a pinned support at point A, a roller support at point D, and a concentrated load applied at point B.

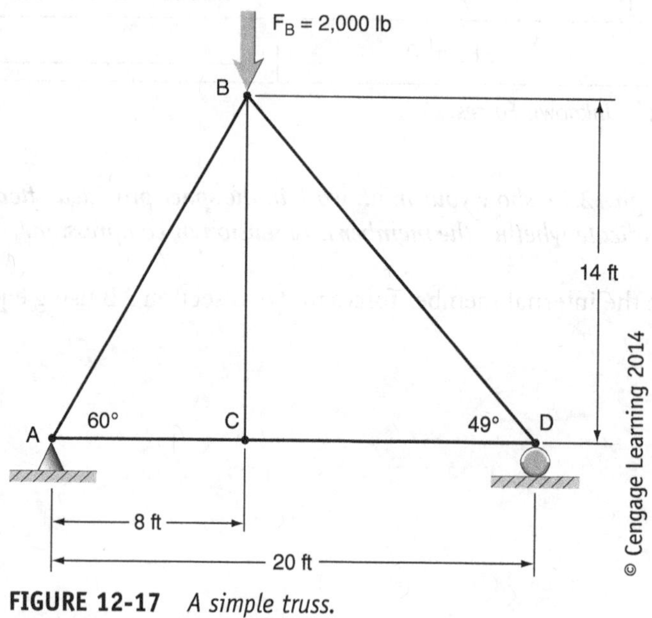

FIGURE 12-17 *A simple truss.*

1. In the space provided, draw and label a free-body diagram of the entire truss.

2. Calculate the vertical reaction force at point D by summing the moments about point A ($\Sigma M_A = 0$). Show your math work and record your answer in the space provided.

R_{DY} = ______________________ lb

3. Calculate the vertical reaction force at point A by summing the forces in the *y*-direction ($\Sigma F_Y = 0$). Show your math work and record your answer in the space provided.

R_{AY} = ______________________ lb

4. Calculate the horizontal reaction force at point A by summing the forces in the *x*-direction ($\Sigma F_X = 0$). Show your math work and record your answer in the space provided.

R_{AX} = ______________________ lb

5. In the space provided, draw and label a free-body diagram of each joint showing all forces acting on and within the truss. *For consistency, assume all members to be in tension, drawing arrows (F_{AB}, F_{AC}, F_{BD}, F_{CD}, and F_{BC}) pointing away from the joints. For reference, see Figure 12-49 on page 422 of your textbook.*

For Problems 12.16 through 12.20, show your math work in the space provided after each problem. Record your answers in Table 12.3 and indicate whether the member is in tension or compression.

Problem 12.17 Calculate the internal member forces for truss section AB from joint A using equilibrium equation $\Sigma F_{AY} = 0$.

Problem 12.18 Calculate the internal member forces for truss section AC from joint A using equilibrium equation $\Sigma F_{AX} = 0$.

Problem 12.19 Calculate the internal member forces for truss section BD from joint D using equilibrium equation $\Sigma F_{DY} = 0$.

Problem 12.20 Calculate the internal member forces for truss section CD from joint D using equilibrium equation $\Sigma F_{DX} = 0$.

Problem 12.21 Calculate the internal member forces for truss section BC from joint C using equilibrium equation $\Sigma F_{CY} = 0$.

Truss Member	Internal Force (lb)	Tension/Compression
AB	F_{AB} = ____________	____________
AC	F_{AC} = ____________	____________
BD	F_{BD} = ____________	____________
CD	F_{CD} = ____________	____________
BC	F_{BC} = ____________	____________

TABLE 12.3 *Truss Member Internal Forces.*

Explore Your World

The failure of a structure to withstand the conditions and elements for which it was designed can have devastating consequences. Whether it's the failure of a bridge to support a load, a dam or levy to contain or control a body of water, or a carnival ride to provide adequate restraint for thrill-seekers, the study and analysis of design failures can lead to new innovations in materials, construction methods, design parameters, and safety regulations. *To Engineer Is Human: The Role of Failure in Successful Design* (Vintage Books, 1992), a book by Henry Petroski, explores a handful of design failures, most notably the collapse of the suspended walkway at the Hyatt Regency hotel in Kansas City, which killed 114 people and injured nearly 200 others. An analysis of the walkway's structural inadequacies provides a relevant context to apply what you have learned in this chapter about the nature of forces.

The original design for the walkway support system (shown in Figure 12-18a) called for a set of continuous tie rods to connect the second- and fourth-floor walkways and suspend them from the ceiling, with the second-floor walkway hanging directly below the fourth-floor walkway. To better understand this configuration, imagine for a moment that you are hanging from a rope that is suspended from a tree branch. A friend of equal weight is holding onto the same rope some distance below you. You are both holding onto the rope with only your hands and transferring your weight to the rope through your grips.

For a variety of reasons, the suspended walkways were *not* constructed according to the original design. Rather than using a set of continuous rods to connect both walkways to the ceiling, one set of tie rods was used to connect the second-floor walkway to the fourth-floor walkway and a separate set of tie rods was used to suspend the fourth-floor walkway from the ceiling (as shown in Figure 12-18b). Let's apply a similar design modification to the way that you and your friend are hanging from the tree. Imagine that your rope isn't long enough for you both to hang from the branch using the same rope. Instead, you have two shorter ropes that will effectively accomplish the same objective. In the new scenario, you hold onto one rope that is connected to the tree branch, while your friend hangs from a second rope that is tied to your leg. It does satisfy the same goal, in that you are suspended in tandem from the tree branch. However, the modified design has altered the way in which your friend's weight is now transferred to the rope that's attached to the tree branch. How do you think this new configuration affects the amount of weight that is supported by your grip on the rope?

To understand *how* changes to the design of the walkway's box beam and rod support system came about, and *why* the design changes led to the failure of the suspended walkway system, do the following:

- Read Chapter 8, "Accidents Waiting to Happen," in the Petroski book, and/or conduct an Internet search using the keywords *Hyatt Regency Walkway Case Study.*
- Study the schematics of the fourth-floor box beam support system as it was originally designed in Figure 12-18a, and as it was redesigned and constructed in Figure 12-18b. Note that P represents the load on each of the components of the support system.

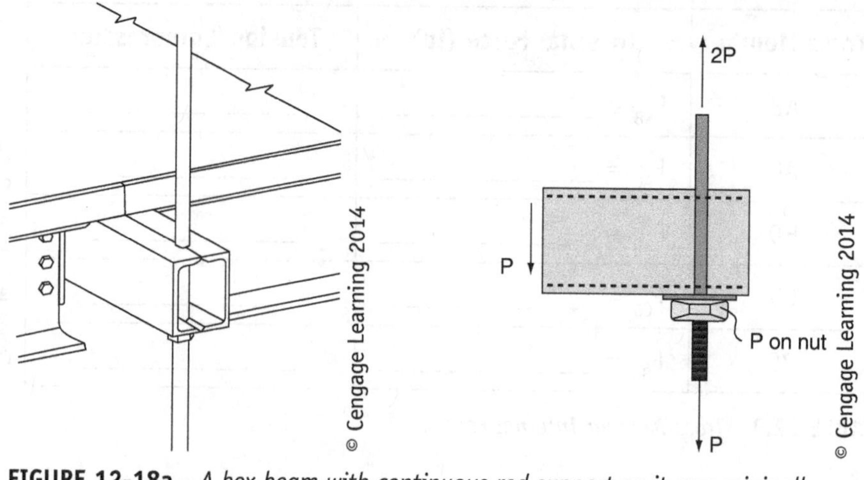

FIGURE 12-18a *A box beam with continuous rod support as it was originally designed for the fourth-floor walkway.*

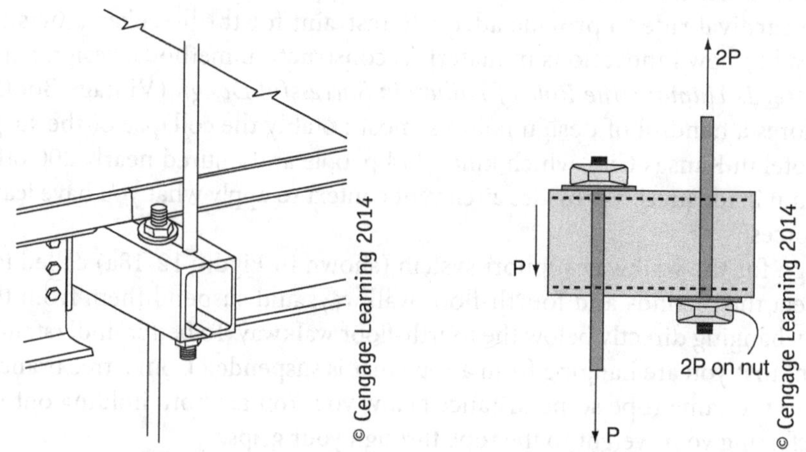

FIGURE 12-18b *A box beam with staggered rod support as it was constructed for the fourth-floor walkway.*

After concluding your research on the collapse of the Kansas City Hyatt Regency walkway, explain in your own words why you think the box beam system in Figure 12-18b failed to support the load of the patrons.

CHAPTER 13
Kinematics and Trajectory Motion

Before You Begin

Think about these questions as you study the concepts in this chapter.

- What are the basic kinematic terms?
- How does a vector quantity differ from a scalar quantity?
- Why does a projectile take a parabolic flight path?
- In what everyday applications can kinematics be used?

Classical Mechanics

Explore Your World

Mechanics is a branch of applied physics, the roots of which can be traced back to the writings of the ancient Greek philosophers. It is the study of motion, forces, and the effects of forces on objects. The three scientists credited with what is now known as *classical mechanics* are Galileo Galilei, Johannes Kepler, and most notably, Sir Isaac Newton.

It is important to gain a historical perspective of the conditions and the thinking that prevailed during the times that these early innovators lived. Such perspective can provide some meaningful context to help us question, investigate, and engineer within our physical world.

Start your exploration of this topic by doing the following:

- Form a team of three students, in which each student will conduct an independent Internet search on one of the three scientists: Galileo, Kepler, or Newton. Make sure that each student has selected a different person. Each student will compose a one- to two-page report documenting the contributions that his or her scientist made to classical mechanics.
- Work with your team members to combine your individual reports into a Microsoft® PowerPoint® presentation that gives an accurate, chronological history of the progress made in the field of mechanics. The presentation should illustrate how each scientist's research contributed to the cumulative body of knowledge of mechanics and advanced the discipline over their lifetimes.

Basic Kinematic Terms

Exercise 13.1 Vocabulary Application

Objective

At the conclusion of this exercise, you will be able to do the following:

1. Identify scalar and vector quantities.
2. Explain the difference between key kinematic terms.
3. Apply the correct kinematic term when given a contextual statement.

Procedure

Read the section on basic kinematic terms (pp. 428–430) in Chapter 13, "Kinematics and Trajectory Motion," of your *Principles of Engineering* textbook. Conduct additional research on the Internet, using each kinematic term in a keyword search.

1. Complete the following vocabulary exercises, and record your answers in the spaces provided.

   a. Label each of the following kinematic terms as either a *vector* quantity or a *scalar* quantity.

   A. Speed ____________________

   B. Acceleration ____________________

   C. Distance ____________________

   D. Velocity ____________________

   E. Displacement ____________________

   F. Gravitational acceleration ____________________

b. Explain in your own words the difference between the use of the terms *distance* and *displacement* as they apply to physics. Be specific and provide examples to highlight the contrast.

____

____

____

____

____

c. Explain in your own words the difference between the use of the terms *speed* and *velocity* as they apply to physics. Be specific and provide examples to highlight the contrast.

____

____

____

____

____

d. Read each of the sentences below. Choose the kinematic term (A – F) from Exercise 13.1a that corresponds with the quantity described in each sentence. Place the letter of the term next to the sentence in the space provided. Some terms may be used more than once.

_________The vertical motion of the batter's ball slowed at this constant rate as it climbed toward the peak of its flight path and then began to speed up at the same rate as it descended toward the outfield.

_________The vehicle's odometer indicated that the car had been driven a total of 50,000 miles.

_________The woman drove her car 70 mph in the eastbound lane of the throughway.

_________During his excursion, the hiker walked 4 miles east, and then turned to follow a stream 3 miles north, leaving him 5 miles from the point where he started.

_________The driver pulled away from the intersection, changing his speed from 0 mph to 60 mph in 4 seconds.

_________The soccer player kicked the ball 40 feet.

_________Using this constant, the prankster was able to determine the approximate time that he should drop the water balloon from the roof in order to hit the moving target below.

_________The cyclist pedaled her bicycle 20 mph.

Analyzing Projectile Motion

Exercise 13.2 Projectile Motion Rudiments

Objective

At the conclusion of this exercise, you will be able to do the following:

1. Determine the initial vertical velocity of a projectile.
2. Determine the initial horizontal velocity of a projectile.
3. Explain why it is necessary to analyze the vertical and horizontal aspects of motion separately.
4. Determine the time that it takes for a projectile to reach the apex of its flight path.

5. Determine the maximum height of a projectile.
6. Determine the total flight time of a projectile.
7. Determine the total range of a projectile.
8. Determine the launch angle of a projectile.

Procedure

Read the section on projectile motion (pp. 430–436) in Chapter 13 of your textbook.

TIP SHEET

As you may recall from Chapter 13, gravitational acceleration moves in a downward direction and is represented as a negative value (-9.8 m/s^2 or -32.2 ft/sec^2). Remember to substitute the number *and* the negative sign when substituting for g!

Figure 13-1 is a vector drawing representing a projectile that has been launched with an initial velocity of 75 m/s at an angle of 55° relative to the level ground. For each of the problems below, show your math work and record your answer in the space provided.

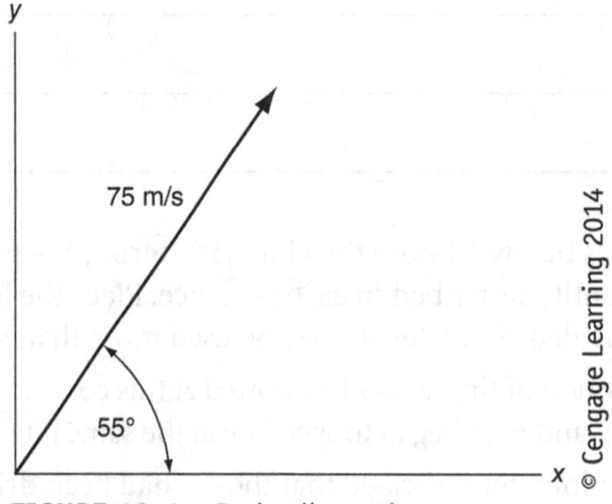

FIGURE 13-1 *Projectile motion.*

Problem 13.1 Calculate the initial *vertical* velocity of the projectile, as illustrated in Figure 13-2. (See Equation 13-2 on page 434 of your textbook.)

V_{iy}= ______________m/s

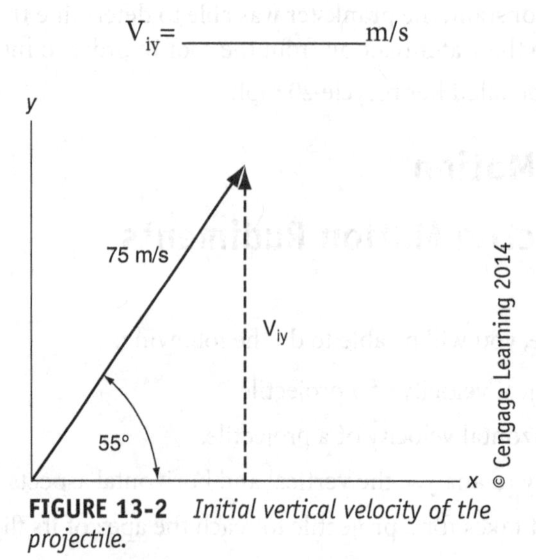

FIGURE 13-2 *Initial vertical velocity of the projectile.*

Problem 13.2 Calculate the initial *horizontal* velocity of the projectile, as illustrated in Figure 13-3. (See Equation 13-1 on page 433 of your textbook.)

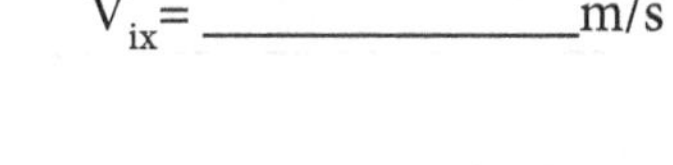

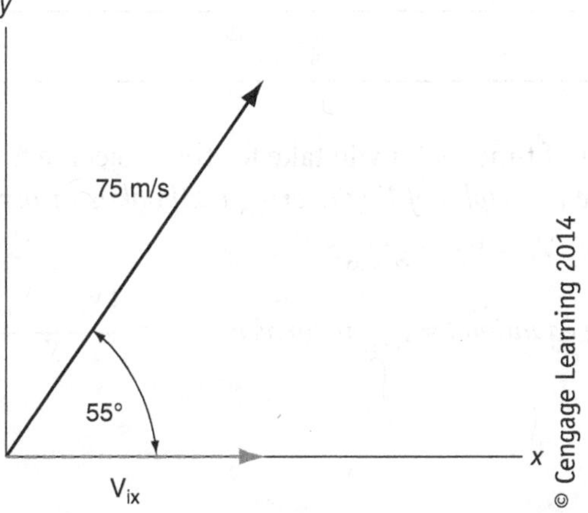

FIGURE 13-3 *Initial horizontal velocity of the projectile.*

Problem 13.3 When solving projectile motion problems, why is it necessary to consider the vertical and horizontal components of a projectile's initial velocity separately?

____

____

____

____

Problem 13.4 What is the vertical velocity (V_y) of the projectile at the very top (apex) of its flight path?

V_y = ____________m/s

Explain your answer:

____

____

____

____

Problem 13.5 What is the horizontal velocity (Vx) of the projectile at the very top (apex) of its flight path?

V_x = ______________m/s

Explain your answer:

____

____

____

____

Problem 13.6 Calculate the amount of time that it will take for the projectile to reach the apex of its trajectory. *Note: Equation 13-3 on page 434 in the Principles of Engineering textbook is incorrect. The correct equation for V_f (velocity at the apex of the flight path) is: $V_f = V_{iy} + gt_{max}$.*

The equation for t_{max} is correct: $t_{max} = \frac{(V_f - V_{iy})}{g}$

t_{max}= ______________s

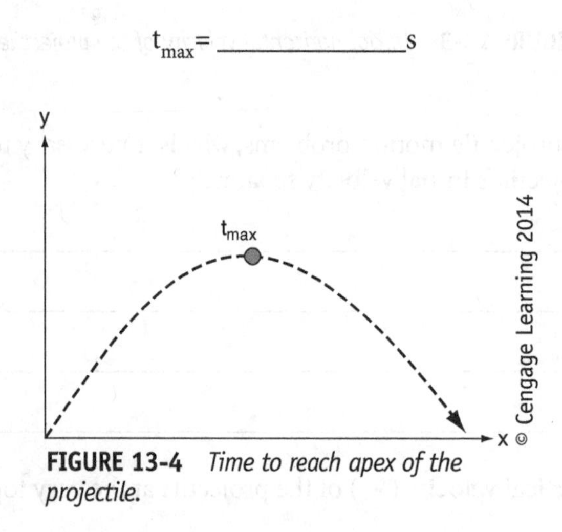

FIGURE 13-4 *Time to reach apex of the projectile.*

Problem 13.7 Calculate the maximum height that the projectile will reach. (See Equation 13-4 on page 435 of your textbook.)

d_y= ______________m

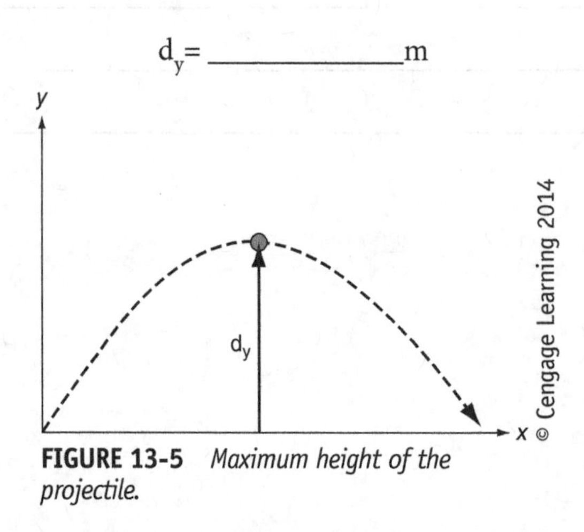

FIGURE 13-5 *Maximum height of the projectile.*

Problem 13.8 Calculate the amount of time that it will take before the projectile will hit the ground.

t_{total} = _______________ s

Problem 13.9 Calculate the maximum range of the projectile. (See Equation 13-5 on page 436 of your textbook.) Caution: Consider which value of time, t_{max} or t_{total}, you should use!

d_x= _______________ m

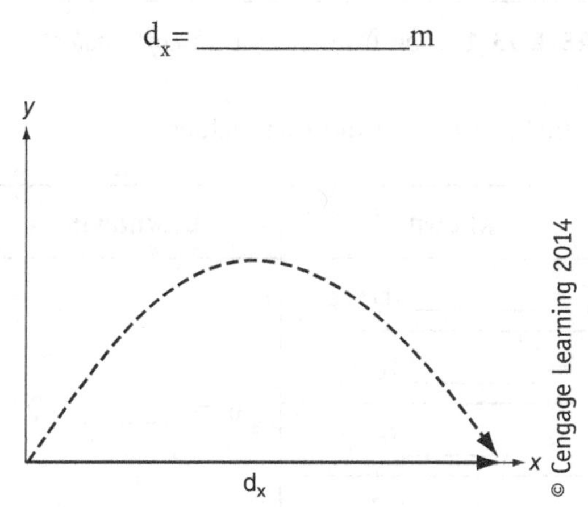

FIGURE 13-6 *Maximum range of the projectile.*

Projectile Motion Application

Exercise 13.3 Projectile Motion Problems

Objective

At the conclusion of this exercise, you will be able to apply kinematic equations to solve projectile motion problems.

Procedure

Read the section on projectile motion (pp. 430–436) in Chapter 13 of your textbook.

A golfer, who has the ability to hit a ball with enough force to produce an initial velocity of 90 ft/sec, must clear a 79-foot tree on the fairway by 1 foot (80-ft maximum height) in order to have the ball land on the green 242 feet away (see Figure 13-7). The golfer may choose from the clubs listed in Table 13-1. Acceleration due to gravity (g) is –32.2 ft/sec^2.

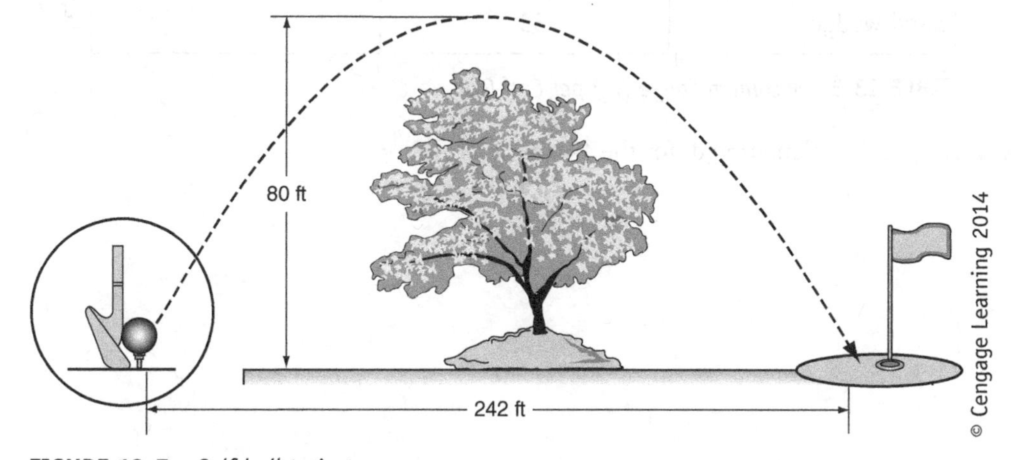

FIGURE 13-7 *Golf ball trajectory.*

Golf Club	Loft Angle
4-iron	25°
6-iron	31°
8-iron	37°
Pitching Wedge	45°
Sand Wedge	53°

TABLE 13-1 *Golf Club Types, with Loft Angles.*

1. Complete Table 13-2 with the known and unknown values.

Known	Unknown
V_i = ________ ft/sec	
d_y = ________ ft	θ = ________ ?
d_x = ________ ft	
g = ________ ft/sec²	

TABLE 13-2 *Kinematic Variables.*

2. Use Equation 13-7 on page 437 of your textbook to calculate the maximum range of the golf ball (d_x) for each of the golf clubs in Table 13-1. Show your work in the spaces provided and record your final results in the d_x column of Table 13-3.

Club	θ	d_x
4-iron	25°	
6-iron	31°	
8-iron	37°	
Pitching wedge	45°	
Sand wedge	53°	

TABLE 13-3 *Maximum Range (d_x) per Golf Club.*

Show work here for calculating d_x for the 4-iron (25°) club:

Show work here for calculating d_x for the 6-iron (31°) club:

Show work here for calculating d_x for the 8-iron (37°) club:

Show work here for calculating d_x for the pitching wedge (45°):

Show work here for calculating d_x for the sand wedge (53°):

3. Which clubs (and which angles) meet the requirement for the maximum range of the golf ball as stated in the question? Record those clubs and their angles here.

______________________ ______________________

4. Of the angles you identified in question 3, which is likely to produce a higher vertical distance? Explain.

____

____

____

____

5. For each of the angles that meet the maximum range requirement, calculate the maximum vertical height, d_y, that the golf ball will reach. To do so, calculate the initial velocity in the vertical direction, V_{iy}, the time to reach the maximum height, t_{max}, and finally, the maximum vertical height, d_y.

   a. Using Equation 13-2 on page 434 of your textbook, calculate the velocity in the vertical direction, V_{iy}, for each of the clubs (and their corresponding angles) that meet the maximum range requirement. In Table 13-4, fill in the information corresponding to the clubs and angles that you are testing, and show your work for each underneath those labels.

Club = ______________ (θ = ________)	**Club = ______________ (θ = ________)**
Show your work here to calculate V_{iv} for this club:	Show your work here to calculate V_{iv} for this club:

TABLE 13-4 *Initial Vertical Velocity.*

b. Using Equation 13-3 on page 434 of your textbook, calculate the time for the golf ball to reach its maximum height, t_{max}, for each of the clubs (and their corresponding angles) that meet the maximum range requirement. In Table 13-5, fill in the information corresponding to the clubs and angles that you are testing, and show your work for each underneath those labels.

Club = ______________ (θ = ________)	Club = ______________ (θ = ________)
Show your work here to calculate t_{max} for this club:	Show your work here to calculate t_{max} for this club:

TABLE 13-5 *Time to Reach the Apex (t_{max}).*

c. Using Equation 13-4 on page 435 of your textbook, calculate the maximum height, d_y, reached by the golf ball for each of the clubs (and their corresponding angles) that meet the maximum range requirement. In Table 13-6, fill in the information corresponding to the clubs and angles you are testing, and show your work for each underneath those labels.

Club = ______________ (θ = ________)	Club = ______________ (θ = ________)
Show your work here to calculate d_y for this club:	Show your work here to calculate d_y for this club:

TABLE 13-6 *Maximum Projectile Height.*

d. Which of the golf clubs not only enables the ball to travel the maximum range requirement, but also can clear the tree?

____

Extra Mile

Exercise 13.4 Design Challenge

Objective

At the conclusion of this design challenge, you will be able to do the following:

1. Apply what you have learned about projectile motion to the design of an automated water balloon launcher.
2. Predict the height that the projectile will reach based on the initial trajectory angle of the device.
3. Predict the horizontal distance the projectile will travel based on the initial trajectory angle of the device.

Procedure

Read Chapter 13 in your textbook.

The *Extra Mile* design challenge for a water balloon launcher is introduced on page 443 of your textbook. The design brief is shown below.

Design Brief	
Problem Statement	In high school, Spirit Week often culminates in a pep rally where upperclassmen engage in friendly competition through games and contests that are designed to entertain and rally the crowd's sense of school pride. The pep rally coordinator is looking for a new idea that combines engineering innovation with a traditional pep rally event.
Design Statement	Design a device that will automate the launching of water balloons during a water balloon toss event and that has the capability of hitting a target at a specified range.
Constraints	The device must conform to the following specifications:
	1. It is easy to assemble and disassemble.
	2. It uses a mechanical form of energy transfer (air pressure and chemical combustion are strictly forbidden).
	3. It can launch a balloon at a uniform initial velocity at any angle.
	4. It must be able to adjust the water balloon's initial trajectory angle between 45° and 90°.
	5. It must be able to hit a target at a specified range.

When meeting with your design team, consider using the following prompts to help initiate a discussion related to the functional aspects of the water balloon launcher. Have each team member share her or his theory about each question, then complete the following exercise to help guide the team's design process.

Why might the device be limited to trajectory angles between 45° and 90° (as specified in constraint #4)? Why not be able to adjust the launch angle to less than 45°?

Why would it be desirable for the device to be capable of launching the projectile (water balloon) with a uniform initial velocity for all launch angles (as specified in constraint #3)?

What are the features of your water balloon launcher that will affect:

- the maximum height that the water balloon will reach?
- the maximum horizontal distance the water balloon will travel?
- the amount of time that it takes for the water balloon to reach its intended target?

Figure 13-8 represents the parabolic flight path of a water balloon launched with an initial velocity of 22.36 ft/sec^2.

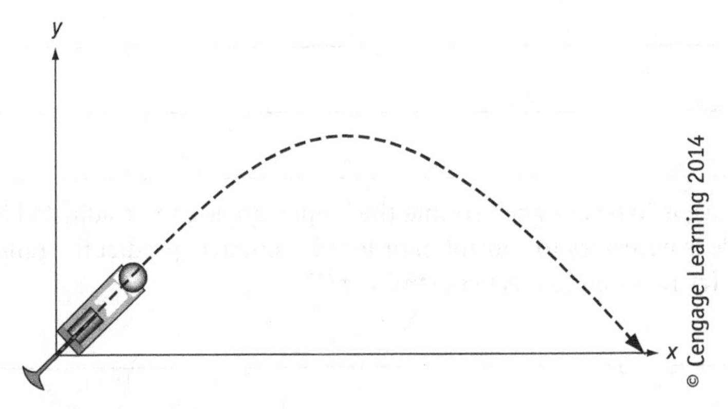

FIGURE 13-8 *Water balloon launcher.*

1. Calculate the range of the projectile (total horizontal distance) when it is launched at angles of 5°, 15°, 25°, 35°, 45°, 55°, 65°, 75°, and 85°. Use a value of −32.2 ft/sec^2 for acceleration due to gravity. Round your answers to the nearest foot. Record your answers in Table 13-7. (See Equation 13-7 on page 437 of your textbook.)

Launch Angle	Range (d_x)
5°	________ ft
15°	________ ft
25°	________ ft
35°	________ ft
45°	________ ft
55°	________ ft
65°	________ ft
75°	________ ft
85°	________ ft

© Cengage Learning 2014

TABLE 13-7 *Maximum Ranges.*

2. Which launch angle produced the longest range?

____

____

3. What have you noticed about the range of the projectile when it was launched at angles above 45°, compared to when it was launched at angles below 45°?

____

____

____

4. Based on the results of the range calculations that you recorded in Table 13-7, explain why you think that it's reasonable to make the device adjustable only for launch angles between 45° and 90°?

____

____

____

5. Given that the device is to be designed so that the launch angle can be adjusted between 45° and 90°, what other variable is necessary to control in order to accurately predict the range of the projectile for each launch angle (as required to meet constraint #5)?

____

____

____

CHAPTER 14
Introduction to Measurement, Statistics, and Quality

Before You Begin

Think about these questions as you study the concepts in this chapter.

- Why is having a standard system of measurement important?
- What are the two most common types of measurement systems used by engineers?
- How are numbers expressed in scientific notation?
- What types of linear measurement tools are commonly used, and how does an engineer decide which measurement tool is appropriate for the task?
- What is the difference between a data set's mean, median, mode, and range?
- What is standard deviation and how is it calculated?
- What statistical methods are used to evaluate quality?

Standard System of Measurement

Explore Your World

In September 1999, after completing a nearly 10-month journey to Mars, a $125 million NASA-commissioned robotic space probe, the *Mars Climate Orbiter,* failed to reach its intended orbit due to a navigational error. An investigation of the mishap revealed that there was a breakdown in communication during the space probe's launch into the orbit of Mars. The flight system software on the *Mars Climate Orbiter* calculated thrust performance using metric units (Newtons) while the ground crew was entering course correction and thruster data using U.S. customary units (pound-force). This discrepancy in units resulted in a miscalculation of the craft's trajectory and caused the orbiter to encounter Mars at an altitude that was lower than anticipated, which is suspected to have caused the orbiter to disintegrate from atmospheric stresses. This is one example, among several that have occurred throughout history, that illustrates the potentially detrimental impact of using disparate measurement systems in an increasingly global society.

- Conduct independent research to find an additional example of how a mismatch in units, or failure to consistently apply a standard of measurement, resulted in disparate data and the inability of two systems to integrate. Use at least two independent websites that describe the example. Some examples to consider include the Space Mountain ride at Disneyland Tokyo, Air Canada Flight 143, and the escape of a 250-kg tortoise.
- Write a brief report that identifies the period of history in which the incident occurred and describes the discrepancy in measurement systems and the failure of integration that resulted. Include the URLs from the independent websites you consulted at the end of the report.

Exercise 14.1 Measurement Conversion

Objective

At the conclusion of this exercise, you will be able to do the following:

1. Convert standard decimal notation to scientific notation.
2. Convert scientific notation to standard decimal notation.
3. Convert fractions to decimals.

Procedure

Read the section on scientific notation in Chapter 14, "Introduction to Measurement, Statistics, and Quality" in the *Principles of Engineering* textbook.

1. Convert the following sets of numbers from their original format to the format specified. Enter your conversions in the spaces provided.
   a. Express the following numbers in scientific notation. Use the convention ($a \times 10^b$), where a is a value greater than 1 and less than 10. For example, the number 4,200 would be written as 4.2 × 103 rather than 42 × 102.

      3,502,000 = ____________________

      0.00152 = ____________________

      14,300,000 = ____________________

      0.000062 = ____________________

      15.0036 = ____________________

      0.0000006325 = ____________________

b. Express the following numbers in standard decimal notation.

5.68×10^{9} = ____________

1.385×10^{-12} = ____________

9.15×10^{-8} = ____________

1.03×10^{6} = ____________

4.86×10^{-7} = ____________

8.043×10^{13} = ____________

c. Convert each fraction in Table 14-1 to its decimal equivalent and enter it into the corresponding location in the table (write the full decimal value). Then convert each decimal into scientific notation and enter it into the corresponding location in the table. Use the same convention for scientific notation as described previously in Exercise 14-1a.

Fraction	Decimal	Scientific Notation
1/16	= ____________	= ____________
1/8	= ____________	= ____________
3/16	= ____________	= ____________
1/4	= ____________	= ____________
5/16	= ____________	= ____________
3/8	= ____________	= ____________
7/16	= ____________	= ____________
1/2	= ____________	= ____________
9/16	= ____________	= ____________
5/8	= ____________	= ____________
11/16	= ____________	= ____________
3/4	= ____________	= ____________
13/16	= ____________	= ____________
7/8	= ____________	= ____________
15/16	= ____________	= ____________

TABLE 14-1 *Fraction to Decimal to Scientific Notation Conversion.*

Measuring Tools and Techniques

Explore Your World

The dial caliper is a precision measuring device that is also very versatile. This single instrument is capable of making four unique types of measurements, as illustrated in Figure 14-10 on page 452 in your textbook. Each feature of the dial caliper has a specific purpose for measuring objects of varying attributes.

- Find an object or multiple objects in your classroom that have dimensions that could be measured with each of the four features on a dial caliper: (1) step distance, (2) hole depth, (3) inside length and diameter, and (4) outside length and diameter. For each of the four caliper features illustrated below, provide three pieces of information related to using that particular feature to measure an object: (1) the name of the object (e.g., coffee mug); (2) a description of the specific dimension of the object to be measured (e.g., the depth of the inside of the cup); and (3) a justification of how this caliper feature helps to measure this specific dimension of this object (e.g., explaining how a particular caliper feature helps to measure the depth of the inside of the coffee mug).

Object: ____________________

Dimension description: ____________________

____

Justification: ____________________

____

FEATURE A: *Step Distance.*

Object: ____________________

Dimension description: ____________________

____

Justification: ____________________

____

FEATURE B: *Hole Depth.*

Object: ____________________

Dimension description: ____________________

____

Justification: ____________________

____

FEATURE C: *Inside Length and Diameter.*

Object: ____________________

Dimension description: ____________________

____

Justification: ____________________

____

FEATURE D: *Outside Length and Diameter.*

Exercise 14.2 Precision Measuring Tools

Objective

At the conclusion of this exercise, you will be able to do the following:

1. Interpret the readout on a dial caliper to the nearest .001 in.
2. Interpret the readout on a micrometer to the nearest .001 in.

Procedure

Read the section on measuring tools and techniques in Chapter 14 of your textbook.

Figures 14-1a through 14-1f are precision measuring instruments (dial calipers and micrometers) that are set to specific measurements. Read the measurement setting on each instrument to the nearest 0.001 in. and record the reading in the corresponding spaces.

FIGURE 14-1a *Dial caliper #1.*

1. Dial caliper #1: ____________ in.

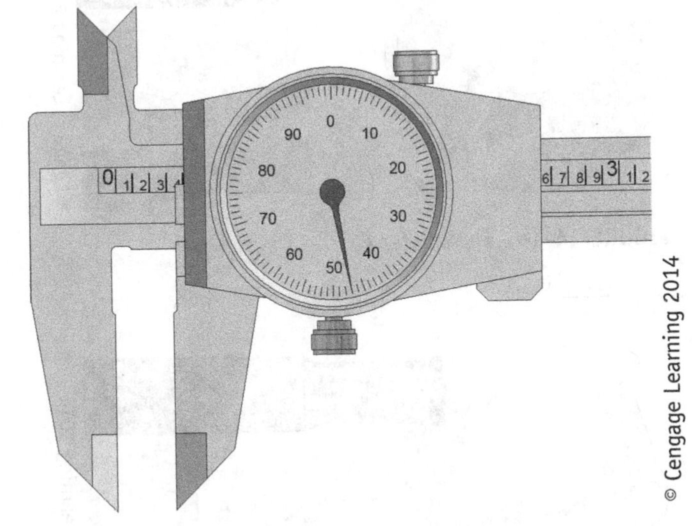

FIGURE 14-1b *Dial caliper #2.*

2. Dial caliper #2: ____________ in.

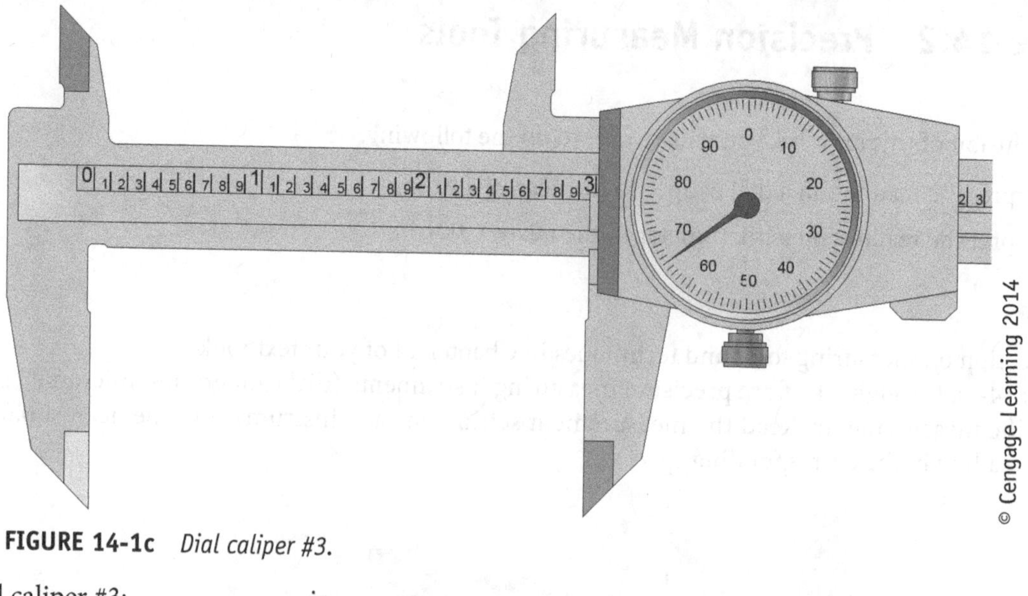

FIGURE 14-1c *Dial caliper #3.*

3. Dial caliper #3: ____________ in.

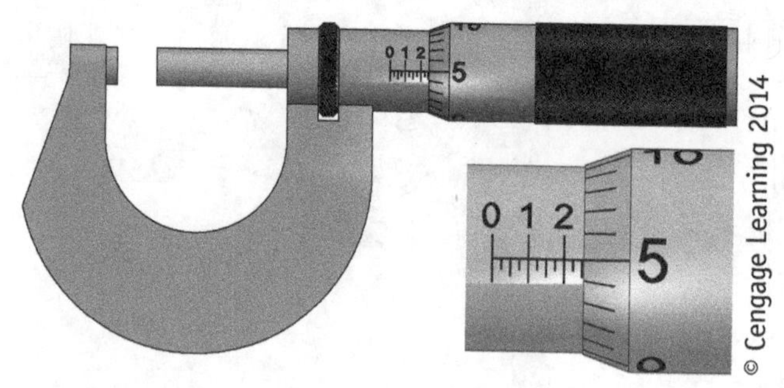

FIGURE 14-1d *Micrometer #1.*

4. Micrometer #1: ____________ in.

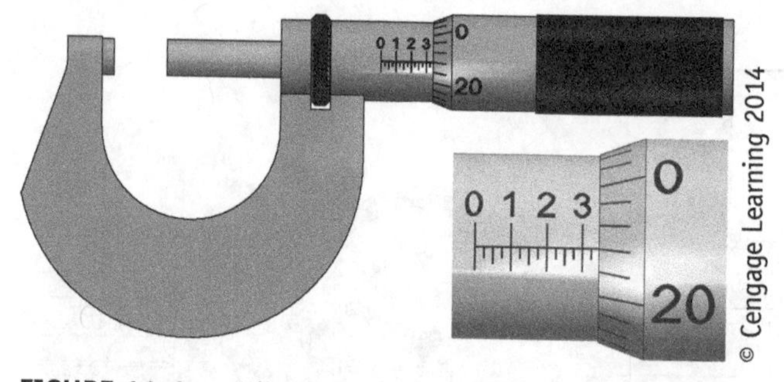

FIGURE 14-1e *Micrometer #2.*

5. Micrometer #2: ____________ in.

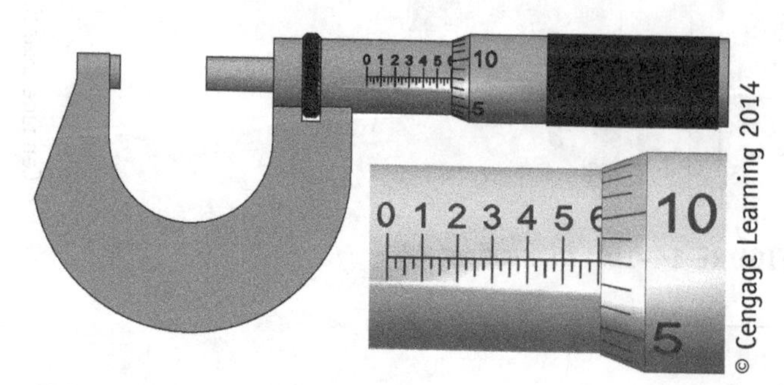

FIGURE 14-1f *Micrometer #3.*

6. Micrometer #3: ____________ in.

Statistical Analysis of Data

Explore Your World

The majority of basic statistical analysis of data involves identifying or calculating the central tendencies of a dataset. The three basic central tendencies of a dataset are the mean, median, and mode. Each of these measures is different, and each has a unique way of describing data. Therefore, each is applied according to the type of data collected and the circumstances under which the data is to be analyzed. For example, if a real estate agent were to characterize home prices in a particular neighborhood to a potential buyer, she would likely use the *median* value of homes in that area. Median values are often used when there are a few extreme values that could influence the mean, or *average*. Since it's typical for houses within a neighborhood to be about the same age, the same size, and on similar pieces of property, they tend to be valued within a certain range. However, there may be some outliers that could skew the mean home value for that area, such as houses that are valued much higher than the rest. The median value, which is the middle value for all the houses in the area, would be more representative of the typical house price.

1. Conduct additional research on the three central tendencies of data (mean, median, and mode) to learn more about the appropriate use of each in analyzing data. Try typing the keywords "when to use… (*substitute mean, median or mode*)" into an Internet search engine.
2. Find one example of a dataset where it would be most appropriate to describe the central tendency of the data using the *mean* value.
3. Find one example of a dataset where it would be most appropriate to describe the central tendency of the data using the *median* value.
4. Find one example of a dataset where it would be most appropriate to describe the central tendency of the data using the *mode* value.
5. Record the three examples in your notebook, making sure to describe the type of data in each example and the circumstances under which it is to be analyzed.

Exercise 14.3 Mean, Median, Mode, and Range

Objective

At the conclusion of this exercise, you will be able to do the following:

1. State an appropriate statistical application for each of the central tendencies (mean, median, and mode) when analyzing data.
2. Determine the mean, median, and mode of a dataset.
3. Determine the range for a given dataset.

Procedure

Read the section on basic statistical analysis of data in Chapter 14 of your textbook.

Table 14-2 contains four sets of data: rows A, B, C, and D. Each row consists of nine samples. Use the information in the table to complete the following exercises. Record your answers in the spaces provided

A	1	13	20	18	4	13	8	12	5
B	20	4	8	17	13	4	13	18	12
C	25	12	21	30	10	28	21	17	21
D	6	1	14	8	2	15	5	1	3

TABLE 14-2 *Sample Datasets A, B, C, and D.*

1. Determine the *mode* (or modes) of each dataset.

A. ______________

B. ______________

C. ______________

D. ______________

2. Which dataset is bimodal, A, B, C, or D? ______________

3. Determine the *range* of each dataset.

A. ______________

B. ______________

C. ______________

D. ______________

4. Which dataset has the greatest range, A, B, C, or D? ______________

5. Determine the *median* of each dataset.

A. ______________

B. ______________

C. ______________

D. ______________

6. Determine the *mean* of each dataset. Round to the nearest (0.0).

A. ______________

B. ______________

C. ______________

D. ______________

Exercise 14.4 Statistical Analysis

Objective

At the conclusion of this exercise, you will be able to do the following:

1. Make comparisons of data based on an analysis of the data's central tendencies.
2. Calculate the standard deviation of a set of data to determine the variability of the data.

Procedure

Read the section on basic statistical analysis of data in Chapter 14 of your textbook.

A process control analyst at a facility that manufactures components for mobile devices was evaluating a set of parts that was produced by the day shift and comparing them to a set of parts that was produced by the night shift. The target thickness of the component was 7 microns (note that 1 micron = 1×10^{-6} meter). An evaluation of 10 samples from each shift produced the results shown in Table 14-3.

	Component Measurements (microns)									
Day Shift:	6.8	6.9	6.9	7.0	7.0	7.0	7.1	7.1	7.2	7.3
Night Shift:	3.7	4.8	5.0	6.1	6.7	6.8	7.8	9.4	9.7	10.0

TABLE 14-3 *Mobile Device Component Measurements.*

Problem 14.1 Comparing Data Based on Central Tendencies

1. After casual inspection of the data, which shift appears to have produced the most accurate parts, the day shift or the night shift? Recall that the target thickness is 7 microns.

____

2. Calculate the *mean* (average) of the dataset from the **day shift**. Show your math work and record your answer in the space provided. Round to the nearest tenth of a micron (0.0).

$\overline{x}$ = __________ microns

3. Calculate the *mean* (average) of the dataset from the **night shift**. Show your math work and record your answer in the space provided. Round to the nearest tenth of a micron.

$\overline{x}$ = __________ microns

4. If you were provided only with the mean value for each dataset, would your conclusion to question 1 be the same? Why, or why not?

____

____

____

5. Calculate the *range* of the dataset from the **day shift**. Show your math work and record your answer in the space provided. Round to the nearest tenth of a micron.

R = __________ microns

6. Calculate the *range* of the dataset from the **night shift**. Show your math work and record your answer in the space provided. Round to the nearest tenth of a micron.

R = __________ microns

7. Which shift had the greatest variability among the parts it produced, the **day shift** or the **night shift**?

____

8. Given your analysis of the data sample from each shift, which component would you prefer to have installed in your mobile device, one that was produced during the **day shift** or one that was produced during the **night shift**? Explain your answer.

____

____

____

____

Problem 14.2 Day-Shift Dataset—Standard Deviation

Use the data from Table 14-3 to calculate the standard deviation in the size of the components that were produced during the day shift.

TIP: The equation for standard deviation is $s = \sqrt{\frac{\Sigma(x - \bar{x})^2}{n-1}}$

Solve for standard deviation by breaking the equation into a series of steps. Record the incremental solutions in Table 14-4.

STEP 1 Start the process by calculating the amount that each value (x_1 through x_{10}) deviates from the mean ($x - \bar{x}$). Record the value of the deviation in the table. Use the mean value ($\bar{x}$) that you calculated in Problem 14-1, 2.

STEP 2 Square each deviation and record the values in the table.

STEP 3 Find the sum of all of the squared values and record the total in Table 14-4. $\Sigma(deviation)^2$ Round to the nearest (0.0).

Day-shift Dataset			
	sample (x)	deviation ($x - \bar{x}$)	(deviation)^2
x_1	6.8	______	______
x_2	6.9	______	______
x_3	6.9	______	______
x_4	7.0	______	______
x_5	7.0	______	______
x_6	7.0	______	______
x_7	7.1	______	______
x_8	7.1	______	______
x_9	7.2	______	______
x_{10}	7.3	______	______
		Σ(deviation)^2	= ______

TABLE 14-4 *Day-Shift Dataset.*

STEP 4 To find the variance, divide the sum (from Step 3) by the number of samples minus 1:

$$\frac{\Sigma(\text{deviation})^2}{n-1}$$

Variance = ______________

STEP 5 Finally, take the square root of the variance (value from Step 4) to find the standard deviation. Round to the nearest (0.0).

s = ______________ microns

1. Use the value for standard deviation that you calculated in Step 5 to fill out Table 14-5 with six standard deviations (±1, ±2, and ±3) from the mean.

−3	−2	−1	$\bar{x}$	+1	+2	+3
______	______	______	7	______	______	______

TABLE 14-5 *Six Standard Deviations.*

2. How many of the components that were manufactured during the day shift fell within:

±1 standard deviation ____________?

±2 standard deviations ____________?

±3 standard deviations ____________?

Problem 14.3 Night-Shift Standard Deviation

Use the data from Table 14-3 to calculate the standard deviation in the size of the components that were produced during the night shift.

Solve for standard deviation by breaking the equation into a series of steps. Record the incremental solutions in Table 14-6.

STEP 1 Start the process by calculating the amount that each value (x_1 through x_{10}) deviates from the mean ($x - \bar{x}$) Record the value of the deviation in the table. *Use the mean value ($\bar{x}$) that you calculated in Problem 14-1, 3.*

STEP 2 Square each deviation and record the values in the table.

STEP 3 Find the sum of all of the squared values and record the total in the table: $\Sigma(deviation)^2$ Round to the nearest (0.0).

	Night-shift Dataset		
	sample (x)	deviation ($x - \bar{x}$)	(deviation)^2
x_1	3.7	______	______
x_2	4.8	______	______
x_3	5.0	______	______
x_4	6.1	______	______
x_5	6.7	______	______
x_6	6.8	______	______
x_7	7.8	______	______
x_8	9.4	______	______
x_9	9.7	______	______
x_{10}	10.0	______	______

⇩

Σ(deviation)^2	= ______

TABLE 14-6 *Night-Shift Dataset.*

STEP 4 To find the variance, divide the sum (from Step 3) by the number of samples minus 1:

$$\text{Variance} = \frac{\Sigma(\text{deviation})^2}{n - 1}$$

Variance = ____________________

STEP 5 Finally, take the square root of the variance (from Step 4) to find the standard deviation. Round to the nearest (0.0).

$s =$ ____________________ microns

1. Use the value for standard deviation that you calculated in Step 5 to fill out Table 14-7 with six standard deviations (±1, ±2, and ±3) from the mean.

−3	−2	−1	$\bar{x}$	+1	+2	+3
______	______	______	7	______	______	______

© Cengage Learning 2014

TABLE 14–7 *Six Standard Deviations.*

2. How many of the components that were manufactured during the night shift fell within:

   ±1 standard deviation ____________?

   ±2 standard deviations ____________?

   ±3 standard deviations ____________?

Problem 14.4 Quality Inspection Report
Create a quality inspection report comparing the day shift and the night shift. For each shift, use the following to describe the shift's performance: mean, standard deviation, and the percentage of components that fell within +/−1, 2, and 3 standard deviations. Be sure to make your comparisons of shift performance and your understanding of standard deviation clear in the report.

Engineering Applications of Statistics

Exercise 14.5 Process Capability

Objective

At the conclusion of this exercise, you will be able to do the following:

1. Use a blueprint tolerance to determine the acceptable range of sizes of a manufactured part.
2. Code the measurements in a sample dataset.
3. Calculate the process capability (Cp) of a machining process.

Procedure

Read the section on engineering applications of statistics (pp. 465–473) in Chapter 14 of your textbook.

A set of dowel pins, shown in Figure 14-2, were manufactured with an overall length specification of 2.625 in. A sample of 35 pins was used to evaluate the manufacturing process. Table 14-8 contains the sample measurements.

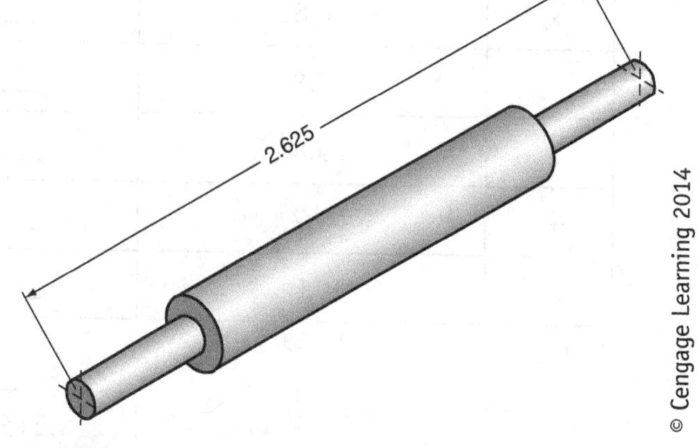

© Cengage Learning 2014

FIGURE 14-2 *Dowel pin sample.*

Dowel Pin Samples (in.)						
2.635	2.621	2.620	2.617	2.626	2.627	2.625
2.631	2.616	2.637	2.626	2.628	2.625	2.623
2.625	2.627	2.630	2.627	2.625	2.631	2.623
2.629	2.630	2.626	2.627	2.620	2.626	2.624
2.631	2.635	2.637	2.618	2.622	2.626	2.630

TABLE 14-8 *Overall Length Measurements for 35 Dowel Pins.*

1. What is the range of acceptable sizes of the dowel pin if the blueprint tolerance for the dimensional length of each pin is ±0.010?

   Size Range: from __________ to __________ in.

   This range constitutes a total blueprint tolerance of __________ in.

2. Code the measurements for the 35 samples and record the coded value for each dimension in the corresponding spaces in Table 14-9.

STEP 1 Create "deviation values": For each measured value, subtract the target dimension (2.625) from the actual measured value, keeping the sign (i.e., if the number minus 2.625 is negative, keep the negative sign).

STEP 2 Change the deviation values to whole value integers by multiplying each deviation value from Step 1 by 1,000. Keep the sign that was determined in Step 1.

Coded Values						
______	______	______	______	______	______	______
______	______	______	______	______	______	______
______	______	______	______	______	______	______
______	______	______	______	______	______	______
______	______	______	______	______	______	______

TABLE 14-9 *Coded Values for Dowel Pin Samples.*

3. How many dowel pin samples fall outside the blueprint tolerance range?

   ____

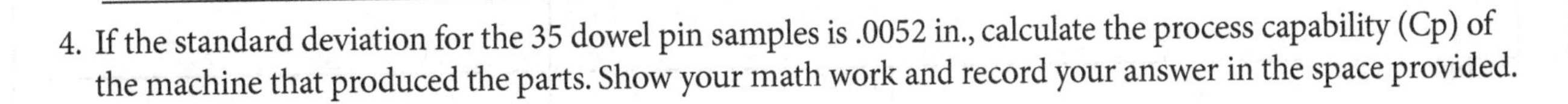

4. If the standard deviation for the 35 dowel pin samples is .0052 in., calculate the process capability (Cp) of the machine that produced the parts. Show your math work and record your answer in the space provided.

   Cp = ____________________

5. Using the table for process capability shown in Figure 14-25 on page 472 of your textbook, how would the manufacturing process that was used to produce the dowel pin samples be characterized?

   ____

Continued from page

Continued to Page

SIGNATURE: DATE:

WITNESSED BY: DATE:

PROPRIETARY INFORMATION

Continued from page

Continued to Page

SIGNATURE:		DATE:
WITNESSED BY:	DATE:	PROPRIETARY INFORMATION

Continued from page

Continued to Page

SIGNATURE:		DATE:
WITNESSED BY:	DATE:	PROPRIETARY INFORMATION

Continued from page

Continued to Page

SIGNATURE:		DATE:
WITNESSED BY:	DATE:	PROPRIETARY INFORMATION

Continued from page

Continued to Page

SIGNATURE:		DATE:
WITNESSED BY:	DATE:	PROPRIETARY INFORMATION